HOMELAND SECURITY
OPERATIONAL ANALYSIS CENTER

T0302668

The Role of Nonprofit Organizations in Community Recovery After Nondeclared Disasters

MELISSA L. FINUCANE, JESSICA WELBURN PAIGE, ANDREW M. PARKER, ANU NARAYANAN,
PEGGY WILCOX, DAVID DESMET, JHACOVA WILLIAMS, KRISTIN VAN ABEL

APPROVED FOR PUBLIC RELEASE; DISTRIBUTION UNLIMITED

This research was published in 2023.

About This Report

The U.S. Department of Homeland Security's (DHS's) Science and Technology Directorate engaged the Homeland Security Operational Analysis Center (HSOAC), a federally funded research and development center (FFRDC) operated by the RAND Corporation for DHS, to help it explore how communities recover from nondeclared disasters. This report provides near-term and longer-term recommendations for DHS to learn from the unique capabilities that nonprofit organizations can bring to disaster recovery. Analyses of six case studies of nondeclared natural disasters examine how nonprofit organizations' social capital provides support particularly to enhance socioeconomic outcomes. This work should be of interest to policymakers and other stakeholders building community resilience to natural disasters in the face of limited federal assistance.

This research was sponsored by DHS's Science and Technology Directorate and conducted within the Infrastructure, Immigration, and Security Operations Program of the RAND Homeland Security Research Division, which operates HSOAC.

About the Homeland Security Operational Analysis Center

The Homeland Security Act of 2002 (Public Law 107-296, § 305, as codified at 6 U.S.C. § 185) authorizes the Secretary of Homeland Security, acting through the Under Secretary for Science and Technology, to establish one or more FFRDCs to provide independent analysis of homeland security issues. The RAND Corporation operates HSOAC as an FFRDC for DHS under contract HSHQDC-16-D-00007.

The HSOAC FFRDC provides the government with independent and objective analyses and advice in core areas important to the department in support of policy development, decision making, alternative approaches, and new ideas on issues of significance. The HSOAC FFRDC also works with and supports other federal, state, local, tribal, and public- and private-sector organizations that make up the homeland security enterprise. The HSOAC FFRDC's research is undertaken by mutual consent with DHS and is organized as a set of discrete tasks. This report presents the results of research and analysis conducted under task order 70RSAT21FR0000148, "Community and Individual Disaster Resilience."

The results presented in this report do not necessarily reflect official DHS opinion or policy. For more information on HSOAC, see https://www.rand.org/hsoac. For more information on this publication, see www.rand.org/t/RRA1770-2.

Acknowledgments

We are deeply grateful to staff within the DHS Science and Technology Directorate who supplied information and insights necessary to conduct this research—especially David Alexander. We sincerely thank interviewees who shared their experiences and perspectives on disaster recovery processes with us. We are also grateful for the comments provided by the reviewers of this report, Cassandra R. Davis of the University of North Carolina at Chapel Hill and Lisa Saum-Manning of the RAND Corporation.

Summary

The Roles and Impacts of Nonprofit Organizations in Long-Term Recovery After Nondeclared Disasters Are Understudied

Disaster recovery is a complex, nonlinear process, characterized by multiple self-organizing actors, including all levels of government, private enterprise, and civil society (Abramson, Grattan, et al., 2015). Nonprofit organizations (NPOs) play a key role in long-term disaster recovery by providing resources or facilitating access to resources (Butler, 2019). One of the main pathways by which NPOs are thought to enable disaster recovery is through social capital—that is, enabling collective action by harnessing the number and quality of networks of relationships among community members (Aldrich, 2012). Although considerable research has examined NPO efforts when federal assistance is available for disaster response and recovery (Curnin and O'Hara, 2019; Eller, Gerber, and Branch, 2015; Simo and Bies, 2007), very few studies have examined NPOs' role in long-term recovery following disasters that do not receive presidential disaster declarations or associated federal resources.[1] The historical presence and continual further integration of NPOs in disaster recovery raise key questions about the nature of the effectiveness of nonprofits as substitutes or complements to governmental and private-sector actors when federal assistance is limited—that is, how are NPOs effective (e.g., are they simply replacing services that other actors would have provided? Or is their support effective for other reasons?)? Whether the benefits of social capital are evenly available to all members of disaster-impacted communities—including residents of underserved communities—is unknown. We define *underserved community* as a population systematically denied full access to economic, social, and civic life and includes populations underserved because of geographic location, religion, sexual orientation, gender identity, minority racial and ethnic populations, or special needs (e.g., language barriers, disabilities, immigration or citizenship status, age) (Biden, 2021; U.S. Code, Title 34, Section 12291).

---

[1] The Disaster Relief and Emergency Assistance Amendments of 1988 (Pub. L. 100-707) amended Public Law 93-288, the Disaster Relief Act of 1974, by (among other things) giving it a new short name: *the Robert T. Stafford Disaster Relief and Emergency Assistance Act*. The 1974 law is codified at U.S. Code, Title 42, Sections 5121 through 5202; collectively, this body of law is commonly referred to as *the Stafford Act*.

The Stafford Act provides two types of disaster declarations: emergency declaration and major disaster declaration. The president can declare an emergency whenever the president determines that federal assistance is needed and can provide up to $5 million in aid. Beyond that amount, Congress must approve the amount. The president can declare a major disaster for any natural event and any fire, flood, or explosion. Amounts and types of aid available under a major disaster declaration are more flexible and varied than under an emergency declaration. For more information on declarations, see Federal Emergency Management Agency (2022a).

Study Objective and Approach

This report addresses three main research questions:

1. What is known about NPOs' role in long-term recovery in the absence of federal disaster assistance?
2. What innovations enhance NPOs' role in underserved communities?
3. What measures best capture whether NPO support for long-term community recovery is effective?

To address these questions, we began by reviewing existing literature on NPOs' role in disaster recovery. We then examined six case studies of nondeclared natural disasters in diverse communities and geographies. We explored publicly available datasets on NPO grantmaking and identified example metrics for assessing NPOs' unique capabilities that are expected to enhance community capacity for socioeconomic recovery. Drawing on the findings from our research, we make near- and longer-term recommendations for collecting and analyzing data to improve understanding of NPO roles and impacts, especially in underserved communities recovering from nondeclared natural disasters.

Key Findings and Recommendations

Overall, this research revealed a lack of empirical evidence about NPOs' role in long-term recovery following a nondeclared disaster. Leveraging social capital is likely an important mechanism by which NPOs can enhance health, social, and economic functioning, particularly in underserved communities, but whether the benefits of social capital are evenly spread is uncertain. Understanding where and how NPOs enhance long-term community recovery requires new approaches to data collection and analysis that support longitudinal, systematic assessments.

Table S.1 summarizes the findings from the research reported in Chapters 2 through 5 and link them to near- and longer-term recommendations that are described in detail in Chapter 6.

Concluding Thoughts

This work provides the U.S. Department of Homeland Security Science and Technology Directorate an assessment of what is known about NPO roles in disaster recovery based on peer-reviewed literature, case studies, and grantmaking data. In short, knowledge gaps are large, and data are sparse. The context-sensitive and evolving nature of disaster recovery means that no single study or data source will be able to answer all questions. Rather, a data system that integrates information across a portfolio of projects with different aims, meth-

TABLE S.1
Linking Findings to Recommendations and Potential Primary Actors

Finding	Term	Recommendation	Stakeholder with a Primary Role in Addressing the Recommendation			
			Nongovernmental		Government	
			NPO	Private Sector	State or Local	Federal
NPOs play a key role in disaster recovery, but the benefits within or across communities might not be evenly distributed, and this particularly disadvantages underserved populations.						
	Near	Improve coordination between government agencies and the nonprofit sector to enhance disaster recovery efforts in underserved communities.	x		x	x
	Longer	Develop guidance (including equity considerations) on how to enhance NPO roles across the disaster cycle.	x			x
Qualitative data are lacking about the mechanisms by which NPOs enhance equitable, long-term economic recovery after nondeclared disasters.						
	Near	To more thoroughly assess NPO roles during recovery, follow up with any community that is denied a presidential disaster declaration.	x	x	x	x
	Longer	Identify a set of communities (including underserved populations) and metrics to track NPO roles and impacts in long-term socioeconomic recovery.	x		x	x
Comprehensive, centralized data on NPO activities in disaster recovery are generally unavailable.						
	Near	Develop a conceptual framework for NPOs' roles in disaster recovery, and use this to prioritize data collection.	x		x	x
	Longer	Foster external partnerships with data-gathering organizations to advance data collection (e.g., determining priorities for survey items) and dissemination.	x	x	x	x
Metrics of NPO effectiveness in disaster recovery need to capture multiple dimensions (socioeconomic outcomes, recovery processes, community social capital, NPO capabilities, and equity in processes and outcomes).						
	Near	Explore options internal to the U.S. Department of Homeland Security for gathering and organizing contextualized information about NPO roles in disaster recovery.	x			x
	Longer	Develop a disaster recovery tracking tool, including metrics related to NPO activities and impacts and community context.	x			x

ods, and time horizons is needed. Moreover, new partnerships and tools are needed to understand and support long-term recovery, especially among the most-underserved members of disaster-affected communities.

Contents

Figures and Tables

Figures

Tables

CHAPTER 1

Introduction

Disaster recovery is a complex, nonlinear process, characterized by multiple self-organizing actors, including all levels of government, private enterprise, and civil society (Abramson, Grattan, et al., 2015). For impacted communities, recovery involves rebuilding social and daily routines and support networks that foster physical and mental health and socioeconomic well-being (Chandra and Acosta, 2009). For a catastrophic event that receives a presidential disaster declaration (PDD), federal assistance mitigates the depletion of state and local monies (due to response costs, falling property values, and lower tax revenues) and reduces the potential decrease in a community's ability to support health care, education, and other social services that affect residents' well-being. Nonprofit organizations (NPOs)[1] also play a key role in long-term disaster recovery by providing or facilitating access to resources (Butler, 2019).

Although most related research has focused on declared disasters, not every community receives approval for a PDD or the associated federal assistance, primarily because FEMA determines states to have adequate resources to manage the recovery process. Very few studies have quantified disaster recovery in non-PDD contexts (Eller, Gerber, and Robinson, 2018), although some studies have investigated the long-term impacts of events for communities trying to support themselves alone. For instance, freedmen's towns, founded between 1865 and 1920, provided superior economic, political, educational, cultural, and other opportunities and agency for Black Americans and help catalyze urban–rural bridging and bonding across generations (Hunter and Robinson, 2018; Roberts and Matos, 2022). Understanding of the direct and indirect roles of NPOs during community recovery is also underdeveloped because most disaster research focuses on immediate response and short-term recovery (up to a month after a disaster) rather than on the long-term recovery phase (from a month to several years after a disaster) (Parker et al., 2020). Thus, little is known about the natural flow of disaster recovery processes and outcomes—and the role of NPOs—when federal disaster assistance is denied, limited, or ad hoc.

Understanding mechanisms of successful disaster recovery is important because the increased frequency, scale, and overlapping nature of disasters is demanding a greater volume and complexity of response and recovery resources (Chandra and Acosta, 2009; National

---

[1] Various terms have been used to refer to institutions that are not profit maximizing, including *voluntary, nonprofit, nongovernmental,* or the *civil society sector.* See Morris, 2000.

Centers for Environmental Information, 2022; O'Brien and Leichenko, 2000). Unfortunately, studies have demonstrated that disasters affect some groups more than others and that current recovery approaches might be exacerbating disparities. Belonging to a certain group—for instance, identified as minority race, low income, female, or younger or older—is related to such outcomes as receiving less disaster funding, having more assistance applications rejected, and suffering more negative impacts on health and economic well-being (Cutter, 2003; Howell and Elliott, 2019; Nelson and Molloy, 2021; Norris et al., 2002; Tierney, 2006; Wisner et al., 2004).

To respond and recover effectively and equitably, communities need to leverage a wide variety of assets and services to support their disaster recovery in a context-sensitive way. Debates about difficult trade-offs need to reflect local needs and values as decisions are made about infrastructure repair, economic development, mitigation, equity and justice, and other issues in the short and long terms (Finucane et al., 2020). Additionally, implementation of recovery plans requires integration of cross-sectoral solutions, which poses logistical (communication and coordination) challenges that are different from those during the disaster response phase (Ballesteros and Gatignon, 2019). For instance, food security depends critically on energy, water, and transportation systems, but, without power for pumps or without roads to access infrastructure needing repair, water might be widely unavailable at first, then unreliable as systems are rebuilt in phases. The historical presence and continual further integration of NPOs in disaster recovery raise key questions about the nature of the effectiveness of nonprofits as substitutes or complements to governmental and private-sector actors when federal assistance is limited, particularly in already-underserved communities. We define *underserved community* as a population systematically denied full access to economic, social, and civic life and includes populations underserved because of geographic location, religion, sexual orientation, gender identity, minority racial and ethnic populations, and special needs (e.g., language barriers, disabilities, immigration or citizenship status, age) (Biden, 2021; U.S. Code, Title 34, Section 12291). A more precise means of assessing NPOs' potential capabilities and performance effectiveness is necessary to establish the likelihood of NPOs' success in disaster recovery (Eller, Gerber, and Robinson, 2018).

Study Research Questions

The U.S. Department of Homeland Security (DHS) Science and Technology Directorate engaged the Homeland Security Operational Analysis Center, a federally funded research and development center operated by the RAND Corporation for DHS, to conduct research to improve understanding of NPOs' role in recovery following nondeclared disasters. This report addresses three main research questions:

1. What is known about NPOs' role in long-term recovery in the absence of federal disaster assistance?
2. What innovations enhance NPOs' role in underserved communities?

3. What measures best capture whether NPO support for long-term community recovery is effective?

Approach

To undertake this study, the research team engaged in multiple research tasks. To inform our understanding of NPOs' role in disaster recovery, we first reviewed literature on select topics, including disaster recovery frameworks, social capital, and recovery metrics related to NPO activities. An initial search identified 244 documents from a variety of sources (e.g., peer-reviewed articles, local and national news, legislative documents, editorials); 238 documents remained after three levels of review (see Appendix A for details).

In a second task, we selected six historic nondeclared natural disasters for case-study analyses:

- tornadoes in
  - Dumas, Arkansas
  - Howard County, Indiana
  - Dallas, Texas
- wildfire in Yarnell, Arizona
- flooding in Tampa, Florida
- climate change (thawing permafrost, flooding, and erosion) in Newtok, Alaska (see Chapter 3 and Appendices B and C for details).[2]

These cases were selected because they represented multiple types of disasters that took place in communities that varied in size, geographic location, and sociodemographic composition. The case-study methods included searching for peer-reviewed literature and other documents describing each of the cases and then interviewing six key informants following a semistructured interview guide. Interview questions focused on disaster impacts, response and recovery, and the roles of the interviewees' organizations. The interviewees included city, county, and state emergency management officials and NPO disaster response coordinators. Interview transcripts were analyzed according to a predetermined coding system. In addition, we searched for publicly available data on NPO grant funding in each of the six case studies (see Chapter 4 and Appendix D for details).

In the third task, we identified recovery metrics (theoretical and extant) for future measurement of NPO effectiveness in supporting long-term community recovery (see Chapter 5).

---

[2] DHS defines natural disasters as

> all types of severe weather, which have the potential to pose a significant threat to human health and safety, property, critical infrastructure, and homeland security. Natural disasters occur both seasonally and without warning, subjecting the nation to frequent periods of insecurity, disruption, and economic loss. (DHS, 2022)

There are important limitations to this study that the reader should keep in mind when considering the key findings and implications of this report. First, the sampled case-study disaster events do not represent the full spectrum of natural hazard types, affected communities, or NPOs across the United States. Thus, case-study findings should not be interpreted as representative of all possible impacts, responses, outcomes, or civil society dynamics. Second, data rely on publicly available media reports, a limited number of interviews, and self-reported grant activities. We have no means of verifying the accuracy of the data or the extent to which data reflect biases or lack of institutional or individual memory about historical events. Third, more systematic detail is available for some cases (e.g., Tampa, Florida; Dallas, Texas) than others (e.g., Dumas, Arkansas; Yarnell, Arizona) because they are better known or more comprehensively studied, in part because of the larger population or geographic areas affected. Fourth, the conclusions we can draw are limited given that federal and state policies have changed over time. For instance, equity considerations have been highlighted in recent years, including through the 2018 Disaster Recovery Reform Act (DRRA) (Pub. L. 115-254, Division D), which allowed the Federal Emergency Management Agency (FEMA) to provide more support to people with more need (previously, everyone was provided with the same amount, regardless of systemic or structural need).

Outline of This Report

After this introductory chapter, Chapter 2 of this report briefly reviews literature aimed at explaining NPOs' role in disaster recovery. Chapter 3 describes the methods and findings from document analysis and interviews related to six case studies of nondeclared disaster events. Chapter 4 describes publicly available grant data that can be used to examine NPO roles and highlights the limitations of the available data. Chapter 5 describes theoretical and existing recovery metrics that might be useful for assessing NPO effectiveness. Chapter 6 summarizes the findings overall and provides recommendations for improving understanding of how community recovery can be enhanced in the absence of federal disaster aid and directions for future research.

This report also includes several appendices. Detailed methods and findings are provided in Appendix A for the case-study literature review and Appendix B for the key-informant interviews. Appendix C provides a descriptive summary of the six case studies. Detailed methods and findings for the nonprofit grant data are provided in Appendix D.

CHAPTER 2

Review of Literature Describing the Role of Nonprofit Organizations in Disaster Recovery

Nonprofit Organizations Bring Unique Experiences and Connections to Support Disaster Recovery

NPOs play a key role in sustained approaches to long-term disaster recovery because of their distinct types of experience and connections with networks of local residents and leaders, regulators and government agencies, and national leaders. NPOs help communities access the information, resources, and expertise needed to tackle complex social challenges in innovative ways (Baregheh, Rowley, and Sambrook, 2009; Dover and Lawrence, 2012; Jaskyte, 2017; McDonald, 2007; Roque, Pijawka, and Wutich, 2020; Sokolowski, 1998). One example innovation is the early adoption of new technologies that might enhance communication with hard-to-reach audiences. Another example innovation is experimentation with new initiatives to meet community needs as they evolve (for instance, an NPO that initially coordinates delivery of food and water supplies might subsequently focus on connecting community members with mental health or social services).

To avoid duplication and bottlenecks, communities need to coordinate with and integrate NPO activities (Chandra and Acosta, 2009; Chandra and Acosta, 2010; Chandra, Acosta, Howard, et al., 2011; P. Joshi, 2010; Moore, Westley, and Brodhead, 2012). For instance, following Hurricanes Katrina and Rita in 2005, NPOs and government entities struggled to coordinate case management services because different entities used different software for tracking clients. Consequently, some residents experienced delays in securing economic, health, or other support or found that those supports were depleted before recovery was achieved. The partnerships among NPOs, between NPOs and communities, and between NPOs and government agencies are thus very important (Muller and Whiteman, 2009).

Because NPOs are often a permanent fixture in the community (Eide, 2010; Kilby, 2008; Telford, Arnold, and Harth, 2004), one of the main pathways by which they enable disaster recovery is through social capital—that is, enabling collective action by harnessing the number and quality of networks of relationships among community members (Aldrich, 2011; Chamlee-Wright and Storr, 2011; A. Joshi and Aoki, 2014; Putnam, 2000). Key elements of social capital include engaging with networks that connect people with social and other resources, establishing civic engagement platforms for equitable participation, encouraging norms of reciprocity, and building trust among various groups with diverse interests (Bhandari and Yasunobu, 2009; Paldam, 2000; Parks et al., 2020; Villalonga-Olives and Kawachi, 2015).

Several types of social capital have been described:

- **Bonding social capital** is characterized by strong ties and assistance between members of a group (e.g., family, friends, neighbors, close allies) and can be driven by culture, religion, ethnicity, or identity. After a disaster, for instance, families and friends will try to connect quickly to determine what material or emotional assistance is needed for members of their immediate community.

- **Bridging social capital** is characterized by connections across different groups with a common goal and is driven by the need for new information (groups can exchange knowledge, experience, and capital). For instance, people from different cultural or socioeconomic backgrounds might share information about how to access disaster assistance, such as connecting people who have lost their home with others who could provide temporary accommodation.

- **Linking social capital** highlights the connections between communities and external (not local) resources, including NPOs, government agencies, other communities, and diaspora relationships. For instance, an NPO might connect with a government agency or other funding organization to access support for disaster recovery activities, such as clearing debris to help rebuild venues for important community services (e.g., schools, health care, utilities) (Aldrich, 2012; Aldrich, Meyer, and Page-Tan, 2018; Behera, 2021; Zhang, 2016).

Leveraging social capital has been associated with better health and well-being outcomes during long-term recovery (Macinko and Starfield, 2001; Putland et al., 2013; Szreter and Woolcock, 2004). Some research suggests that, although households with higher levels of damage experience slower recovery, those with stronger personal networks (larger, denser, closer, longer) and higher levels of social capital (more civic engagement, contact with neighbors, and trust) experience faster recovery (Aldrich, 2011). Other studies, however, suggest that predisaster social capital might have negative consequences for communities. For instance, tight-knit neighbors might prevent unwanted projects, such as temporary trailer parks that are needed to shelter other residents whose homes were damaged in a disaster, thus hindering longer-term housing recovery (Aldrich and Crook, 2008). One explanation for discrepant findings is that certain types of social capital might be more or less beneficial than others in a disaster context and these relationships might differ across groups within an affected community (Parks et al., 2020). In the trailer park situation, the strong bonding social capital among residents in one neighborhood might prevent them from bridging with residents in other neighborhoods who need them to allow in trailers so they can be close to postdisaster rebuilding efforts (Putnam, 2000).

How Nonprofit Organizations' Capabilities Help Build Capacity for Disaster Recovery in Communities

NPOs draw on multiple capabilities to utilize social capital to support innovative processes, services, and products that enable effective disaster response, especially in providing humanitarian assistance. Their unique experiences and connections can reflect long-term vision and commitment, derived from integrative and long-run relationships that afford a deep knowledge of stakeholder dynamics, local needs, cultural values and norms, and institutional logics (Austin, 2010). Specific examples of how NPO capabilities help build capacity for disaster recovery in communities are provided in Table 2.1. For instance, an NPO might help connect multiple neighborhood groups concerned about local flooding with ongoing city- or state-level activities aimed at developing a hazard mitigation plan. Deep trust between local NPOs and communities helps facilitate coordination and communication among community members and between local community groups and nonlocal organizations able to provide recovery resources (e.g., state or federal agencies). NPOs also help com-

TABLE 2.1

Capabilities of Nonprofit Organizations That Enhance Capacity for Disaster Recovery in Communities

Category	NPO Capability	How NPOs Help Build Capacity for Disaster Recovery in Communities
Long-term vision and commitment	Develop strategies for meaningful change.	People in communities are part of the solution rather than the problem.
	Coordinate planning that links small local initiatives into larger (e.g., city-, county-, or state-level) programs.	Community ideas form the foundation for meaningful change.
Cross-sectoral collaboration	Pool resources to achieve better outcomes through effective and efficient collaboration.	Services and initiatives are better aligned with people's daily lives.
	Long-term socioeconomic goals (e.g., access to healthy livelihoods for all) underpin projects and programs.	Socioeconomic goals support the achievement of equitable recovery outcomes (e.g., racial differences in unemployment rates are reduced).
Effective relationship-building	Ensure that involvement is democratic and relationships are respectful.	Local people feel trusted and respected.
	Engage local people in developing positive strategies.	Local people are engaged and have access to support networks.
Generating contextualized knowledge	Resources are available to gather information about what works and why in the local context.	Lessons learned come from community-based participatory processes and contribute to iterative improvement aligned with local needs and values.
	Local knowledge is collected and used to improve practice.	People feel valued and able to work as partners to develop recovery strategies.

munities identify diverse sources of information and data (quantitative and qualitative) that could inform trade-offs in decision making and contextualize interpretation of those data. For instance, understanding experiences of local labor conditions (e.g., historical declines in agriculture-related jobs) could explain generational differences in support for alternative land-use programs. Through their involvement with policy development, enforcement, or advocacy, NPOs bring real-world understanding of barriers and opportunities for implementation (Acosta, Chandra, Towe, et al., 2016b) and help clarify different perspectives and raise awareness of historically underserved communities (Ballesteros and Gatignon, 2019).

Eller, Gerber, and Robinson (2018) describes a unique "intersector operational collaborative model" between voluntary NPOs and state and federal agencies in the context of disaster-scale flooding on the Yukon River in Alaska in 2013. They found that innovative, established, and collaborative arrangements between federal agencies and NPOs can provide long-standing value (e.g., in the context of repeated disasters). In the case of the Alaska disaster described by Eller, Gerber, and Robinson, the use of NPOs to lead the recovery effort promoted community bonding and bridging and provided flexibility through the voluntary nature of the work. The NPOs also helped community members avoid potentially adverse psychological, health, and financial impacts of moving away.

The Benefits of Social Capital Might Not Lead to Equitable Recovery

Although social capital can be positively associated in some instances with various (absolute or aggregate) policy outcome measures, it can be negatively associated with indicators of racial and other social disparities (Andrews, 2012; Hero, 2003a; Hero, 2003b; Hero, 2007). For instance, the finding that better health is found in places with high social capital might apply only to nonminority populations with access to relevant networks. In other words, the benefits of social capital are not necessarily evenly distributed across communities. The positive (or negative) associations between social capital and health, economic, or other outcomes might be true for only some people (Hawes and Rocha, 2011).

One reason for different associations relates to the extent of engagement between NPOs and the private sector or public institutions. As described above, bonding social capital might be strong even where bridging social capital is not, which can ultimately limit the connections with additional external recovery resources. For instance, some NPOs might have strong connections with community members (e.g., an urban farming organization might know which local residents face food insecurity) but have weak connections with private or public funding sources for long-term recovery activities. Conversely, a national or global NPO with employees who live outside a community (e.g., Habitat for Humanity) might lack local ties but be well connected with potential funding partners (Aldrich and Crook, 2008).

A second reason for different associations relates to how well the NPO support matches the needs of communities, local institutions, and their policies or procedures (Rothstein and

Stolle, 2008). For instance, an NPO might provide ways to connect community members with social or health services, but a local government emergency response agency might adhere to a top-down process for assessing and addressing community needs based on methods that prioritize homeowners over renters. A key implication is that NPOs might be ineffective, despite their own capabilities, if they face policies or procedures or other factors that inhibit uptake of NPO support more for some groups than others. Similarly, federal policies or procedures might bolster or dampen NPO effectiveness in supporting equitable recovery outcomes. Although federal assistance is generally assumed to ensure efficiency and coordination during recovery, recent studies have demonstrated that existing policies and processes for distributing that aid in fact contribute to expanding wealth inequalities to homeowners by race over time (Howell and Elliott, 2018; Howell and Elliott, 2019). Mechanisms contributing to expanding inequalities might include differential access to government assistance or uneven disruption to housing or income (Markhvida et al., 2020; Rivera, Jenkins, and Randolph, 2022). Equity considerations need to be incorporated into federal guidance about involvement of NPOs in planning recovery efforts and other initiatives (e.g., relaxing matching-funds requirements to enhance the timeliness of NPOs' access to funds).

Practical Guidance About Partnering with Nonprofit Organizations in Disaster Recovery Is Limited

A social capital approach draws attention to the connections between multiple layers of actors involved in disaster recovery. The interdependence of various levels of decision making and action—policy, practice, and community—underscores the joint responsibility of all levels of government agencies and nongovernmental organizations (NGOs) for successful implementation of disaster recovery initiatives that lead to equitable health and well-being outcomes.

Because disaster recovery depends on interdependent actors at multiple levels, successful outcomes depend critically on the match between NPO capabilities and community capacity to leverage those capabilities. In Table 2.2, we describe some conditions (that need to be supported primarily by local governments) that influence whether NPOs can contribute to successful disaster recovery.

Some policy documents call for improved partnerships, both among and across NPOs (FEMA, 2011a; FEMA, 2011b). For instance, the National Disaster Recovery Framework highlights the importance of strategic partnerships with NPOs and identifies a wide variety of stakeholders from local, state, territorial, tribal, and federal government; community-based organizations; business; and academia. National Voluntary Organizations Active in Disaster (National VOAD) is explicitly named in the National Response Framework with a charge to coordinate efforts of various nonprofits across the disaster cycle. There is little guidance provided, however, on how to implement such coordination effectively, particularly during long-term disaster recovery.

TABLE 2.2

Dependencies Between Nonprofit Organizations and Communities for Successful Disaster Recovery

NPO Capability	Required Mechanisms and Support
Access and use funding	Rapid resource transfer between government agencies and NPOs when possible (e.g., federal or local government programs could eliminate cost-share requirements)
Collect and use local data about dynamics, needs, resource availability, and constraints	Generating, maintaining, and ensuring easy access to data (e.g., city governments could support initiatives to maintain information infrastructure)
Integrate into existing stakeholder networks and form new relationships quickly as needed	Routine stakeholder engagement, including forums that NPOs can join as needed (e.g., city governments could support regular in-person and virtual opportunities for discussions across community groups)
Communicate with and organize diverse actors	Reliable communication infrastructure (e.g., local and federal government agencies work with private companies to ensure telecommunication or internet connectivity)
Stay engaged with the community beyond the immediate response period	Creating and updating plans that outline goals for recovery and beyond (e.g., NPOs and city government agencies collaborate on planning activities)

An operational model for integrating NGOs (Acosta and Chandra, 2013) emphasizes that, although NPOs need to be integrated in recovery planning and processes, government agencies have not established the necessary mechanisms for formal partnerships, which limits involvement of NPOs in emergency planning and exercises. The National Disaster Recovery Framework suggests that NPO leaders participate in predisaster planning and postdisaster recovery via a checklist of recovery responsibilities, but even collaborations of NPOs with expertise in disaster response (e.g., National VOAD) face coordination challenges (Executive Office of the President, 2006). Moreover, NPOs situated in marginalized communities are likely to be additionally disenfranchised and potentially disconnected from the government agencies responsible for disaster response and recovery. Questions remain about communities' ability to use NPOs' capabilities to restore routine community functioning (Adie, 2001; Bankoff and Hilhorst, 2009; Runyan, 2006), especially in the context of long-term recovery with limited federal assistance.

Gaps in Knowledge

The literature reviewed above suggests an important role for civil society in disaster recovery, but empirical findings are limited, particularly regarding NPOs' role in long-term economic recovery after nondeclared disaster events. Social capital is considered a key pathway by which NPOs serve to provide financial, physical, emotional, and logistical support and resources. However, the focus of most studies has tended to be on individual- or household-level, rather than community-level, processes and effects. Studies also tend to be descriptive rather than prospectively assessing change in NPO roles or dynamics through the disaster cycle (Acosta and Chandra, 2013; Islam and Walkerden, 2015). Additionally, heavy reliance on cross-sectional data over time limits understanding of how dynamic relationships among various actors in disaster recovery evolve and what effects those relationships might have, especially when the benefits of social capital are unevenly spread across communities (Fafchamps, 2006).

Empirical documentation of the benefits of social capital is complicated by the presence of negative and positive externalities and by the existence of leadership effects and group effects. Moreover, literature suggests an important role for the federal government in disaster recovery, and it is unclear whether NPOs' roles and effectiveness change when there is limited federal assistance. There is a lack of systematic assessment of whether or how, *in practice*, NPOs contribute meaningfully to sustained economic recovery. Thus, gaps remain in knowledge about the processes through which the strategic and social benefits of NPO engagement during disaster recovery can meet the large-scale social challenges being addressed.

Summary of Key Findings

NPOs play a key role in disaster recovery helping communities access information, resources, and expertise through networks of relationships that provide unique understanding of local needs, norms, and dynamics. Some empirical research suggests that innovative, established collaborations between NPOs and public agencies or the private sector help facilitate disaster recovery. Other research suggests that the benefits of NPO networks within or across communities might not be evenly distributed, potentially exacerbating social inequities. Overall, there is limited robust research and practical guidance on how NPOs effectively support community recovery following nondeclared disasters.

CHAPTER 3

Case Studies of Recovery After Nondeclared Disasters

To examine NPOs' role in recovery processes after nondeclared disaster events, we identified six case studies. Figure 3.1 depicts a timeline of the cases, along with federal disaster legislation. The case studies occur across a 15-year period during which new federal legislation was introduced to improve disaster management by, for instance, enhancing collaborations with the nonprofit sector (through the Post-Katrina Emergency Management Reform Act [Pub. L. 109-295, 2006, Title VI]); improving preparation and response to expediting the provision of assistance to NPOs (through the Sandy Recovery Improvement Act [Pub. L. 113-2, 2013, Division B); and raising attention to predisaster mitigation and equity issues (through the DRRA). A main theme that emerged from the case studies is that NPO roles vary considerably in nondeclared-disaster contexts. Factors that affect NPO activities include who

FIGURE 3.1

Timeline of Disaster Case Studies and Federal Legislation

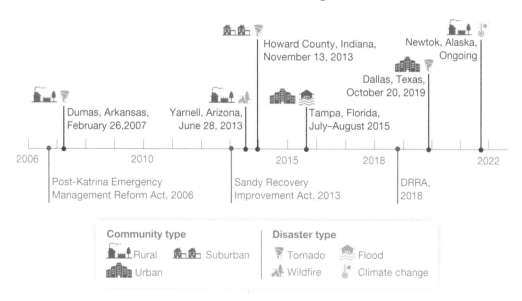

is affected by the disaster, how frequently they experience natural disasters, and the level of coordination between the local government and NPOs.

Approach to Case Studies

We used several criteria to select the cases. First, we focused on communities that had experienced significant natural disasters and been denied a PDD. In each case, a disaster declaration request was made and denied by FEMA. Second, we included a variety of disaster types to gain insight into NPO involvement in different event types. Third, we selected disasters that took place in a diverse set of communities that varied in size, geographic location, and socioeconomic and racial composition. This diversity provided us an opportunity to explore how NPO roles might vary across different types of communities. Fourth, we selected cases that received coverage from media outlets and other sources based our initial scan of available literature to ensure that we could locate information about the community's recovery.

Figure 3.2 maps the case-study locations. A short description of each case is provided in Boxes 3.1 through 3.6.

The variety of cases called for multiple qualitative methods. A literature search of peer-reviewed papers, legislative or other government documents, and media stories yielded 242 documents for analysis (see Appendix A for detailed methods and findings). To gain a better understanding of NPO involvement—including how local governments interact with NPOs after disasters—in-depth interviews were conducted with six key informants from local NPOs or city, county, or state emergency management agencies (see Appendix B for detailed methods and findings).

Nonprofit Organizations' Roles in Recovery Vary After Nondeclared Disaster Events

The document analysis and interviews provided insights about NPO roles across the six case studies (see Table 3.1). The number of NPOs involved ranged from one in the Dumas, Arkansas, case to five in the Yarnell, Arizona, case. The significant NPO response after the Yarnell wildfire seemed to be in part because the deaths of 19 firefighters brought significant attention to the event. In all cases, NPOs focused on providing various types of direct humanitarian relief (e.g., providing food, shelter, or financial assistance). Exceptions were identified in the Tampa, Florida, and Dallas, Texas, cases, in which VOADs' primary role was to coordinate resources and share knowledge across NPOs. Reported monetary values of NPO resources ranged from just over $119,000 to more than $248 million, although figures were not reported for all NPO activities (Anglen and Wiles, 2014; Arizona Community Foundation, 2013; DeFilippis, 2014).

FIGURE 3.2

Map of the Six Case Studies

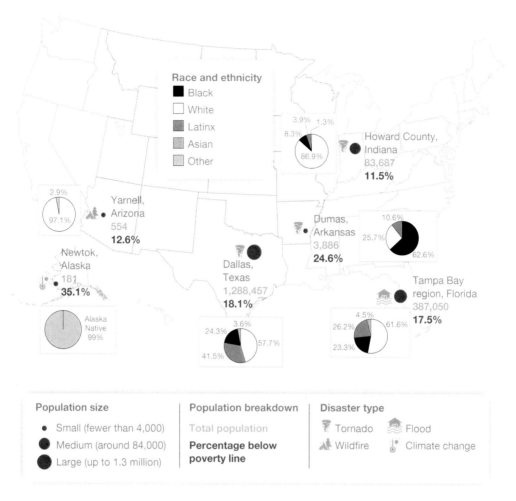

SOURCES: U.S. Census Bureau, undated-a; U.S. Census Bureau, undated-b; U.S. Census Bureau, undated-c; World Population Review, undated-a; World Population Review, undated-b; World Population Review, undated-c.

The nature of NPO activities tended to align with the sociodemographics of the affected communities. In the case of Dallas, Texas, for instance, an NPO provided financial and other assistance for undocumented immigrants who had very few resources to buffer the tornadoes' effects and were less likely than others to seek aid from government agencies. In cases in which multiple or overlapping disasters were experienced (e.g., Howard County, Indiana), NPO resources were described as insufficient to address communities' ongoing and evolving needs.

A common challenge noted across the case studies was that a lack of federal disaster assistance was associated with a slower and more challenging recovery. In Howard County, Indiana, overlapping flood and tornado disasters had exhausted the state's relief funds. In Dallas,

TABLE 3.1

Descriptive Overview of the Roles of Various Nonprofit Organizations Identified for Each Case Study

Case	NPO Role	NPO Identified
Dumas, Arkansas	Direct relief (e.g., provided meals, household items, school supplies)	Catholic Charities USA
Yarnell, Arizona	Direct family financial support; rebuilding homes	100 Club of Arizona
		Phoenix and Prescott Firefighter Associations
		Arizona Community Foundation
		Yavapai County Community Foundation
		Yarnell Memorial Scholarship Endowment
Howard County, Indiana	Direct relief (e.g., bottled water, baby formula) and family financial support	WTHR (local public radio)
		American Red Cross
		United Way
Tampa, Florida	Direct relief (e.g., shelter, food); financial assistance for housing; property repair	American Red Cross
		VOADs
Dallas, Texas	Enabled coordination with local nonprofits; direct financial assistance; support for marginalized people (e.g., undocumented immigrants)	VOADs
		Church of Jesus Christ of Latter-Day Saints
		On the Road Lending
		Catholic Charities USA
Newtok, Alaska	Direct financial assistance for relocation; health services for Indigenous people	Alaska Native Tribal Health Consortium

Texas, the city struggled to meet the cost of repairing local infrastructure, especially public buildings, such as schools. For instance, support to the Dallas Independent School District (Dallas ISD) was provided by a local entrepreneur. In Yarnell, Arizona, federal funding that was not contingent on a disaster declaration was reported to be important for recovery. For example, the U.S. Department of Agriculture Rural Development program provided a $40,000 grant to help the Yarnell Fire District rebuild a helipad destroyed in the fire (Rural Development, 2017). In addition, the U.S. Small Business Administration offered small-business loans of up to $2 million, home loans of up to $200,000, and personal-property loans up to $40,000 (FEMA, 2022b).

The Dumas, Arkansas, Case Study

In February 2007, an Enhanced Fujita (EF) 3 tornado hit the small, rural, predominantly Black town of Dumas, Arkansas.[a] Nearly 25 percent of the town's population live below the poverty line. The tornado injured about 40 residents, destroyed 37 homes, and significantly damaged several local businesses that left 800 residents unemployed. Governor Beebe's request for an emergency disaster declaration was denied because FEMA determined that the state had sufficient resources to manage the response and recovery.

The Yarnell, Arizona, Case Study

In June 2013, a significant wildfire near the small, rural, predominantly White town of Yarnell, Arizona, burned 8,300 acres, destroyed 93 homes, and killed 19 firefighters known as the Granite Mountain Hotshots. About 12.6 percent of the population in this town live below the poverty line. Governor Jan Brewer's request for a major disaster declaration was denied because FEMA determined that the state had sufficient resources to manage the response and recovery. However, significant charitable donations were received to assist families in rebuilding their homes and accessing other resources.

The Howard County, Indiana, Case Study

In November 2013, an EF3 tornado hit Howard County, Indiana, a suburban, predominantly White area with about 11.5 percent of the population living below the poverty line. About 32 residents were injured, thousands remained without power for up to a week, about 60 homes and 30 businesses were destroyed, and homes and infrastructure incurred more than $3 million in damage. Governor Mike Pence's request for a PDD was denied because FEMA determined that the state had sufficient resources to manage the response and recovery.

---

[a] The National Weather Service uses the EF scale to measure the strength of a tornado. The scale ranges from EF0 to EF5. An EF3 tornado is described as severe, with wind speeds between 136 and 165 miles per hour. More information about the scale can be found at National Weather Service (undated).

A county emergency management official candidly explained that, in the case of Tampa, Florida, the fact that federal resources are limited means that not all victims will receive adequate assistance:

> . . . It starts simple with things that you just can't do, like case management. How do you uncover unmet needs if you don't have the money to do case management? And if you don't know what the unmet needs are, how can you even seek out additional funding to help meet those needs, right? So it's like a cycle—there's no winning it, right? Like what few little resources, you're again local—local folks can pour out, and you try to stretch those. But the reality is, some people just won't recover from something like that.

The Tampa, Florida, Case Study

In July and August 2015, three urban counties near Tampa Bay, Florida, received 15 to 20 inches of rain and experienced their worst flooding in 65 years. In this area, about 23 percent of the population is Black and 26 percent is Latinx, and 17.5 percent of the population live below the poverty line. About 1,000 homes were damaged, with an estimated cost of $2.2 million. Governor Rick Scott's request for a PDD was denied because FEMA determined that the state had sufficient resources to manage the response and recovery.

The Dallas, Texas, Case Study

In October 2019, an EF3 tornado hit Dallas, Texas, a large, heavily populated urban area that is nearly 58 percent White, 42 percent Latinx, and 24 percent Black. Damage was estimated at $2 billion, with more than 12 school buildings damaged (three of which had to be demolished). Just over 18 percent of the population lives below the poverty line. Governor Greg Abbott's request for a PDD was denied because FEMA determined that the state had sufficient resources to manage the response and recovery.

The Newtok, Alaska, Case Study

Newtok, Alaska, is a village of 99 percent Alaska Native people, with more than 35 percent living below the poverty line. They have been affected by ongoing flooding, erosion, and thawing permafrost. These climate change–related events have compromised the village's critical infrastructure and made homes unsafe. The primary mitigation effort is to relocate residents to an area about 10 miles away. Village Council president Paul Charles's request for a major disaster declaration in 2016 was denied when FEMA determined that the request fell outside of the Stafford Act's definition of *natural disaster*.

None of the case studies provided evidence of NPO roles in longer-term economic recovery. However, the short-term assistance for property damage (e.g., rebuilding uninsured homes, providing down payments and loans for new vehicles) likely mitigates consumption and income losses that affect houses with limited economic resources more than they affect wealthy households. By providing immediate financial relief, NPOs are helping address the broader well-being losses that can slow or prevent access to employment, education, health care, or food security for already-underserved communities.

A coordinating role for NPOs was apparent in some cases. WTHR, Howard County's local public radio station, published a list of disaster response resources and directed victims to the Red Cross and the United Way of Howard County. In the case of Dallas, Texas, a strong coordination effort between local government and local NPOs reportedly strengthened the disaster response. Similarly, in Tampa, Florida, a high level of coordination between state officials, local governments, and NPOs was considered important, particularly in the absence of federal assistance. A strong state VOAD organization in Florida provides coordination for nonprofits and works closely with the county emergency management office. The VOAD organization also has a representative who works directly with the governor's office, which creates a strong relationship between nonprofits and the state government.

Data Are Very Limited About the Roles of Nonprofit Organizations

The cases we examined consistently highlighted the limited data available about NPOs' roles in recovery following nondeclared disasters. The document analyses and interviews suggest that communities lack consistent methods for tracking recovery outcomes, let alone for tracking NPO roles in long-term recovery processes. For most of our cases, our analysis relied heavily on local and national news coverage of the event, including brief articles and opinion pieces. More-detailed sources, such as policy briefs and academic articles, were limited and often available only for cases that raised an issue considered to be part of a larger national problem, such as shoreline erosion resulting from climate change in Newtok, Alaska, or firefighter deaths in Yarnell, Arizona. In addition, local government turnover can make it challenging to find sources and documents that could provide insight into the impact and recovery. We were able to interview only state and local government officials and NPO staff members with firsthand knowledge of the cases in Dallas, Texas, and Yarnell, Arizona. Officials in Tampa, Florida, were not in their positions at the time of the disaster and had limited information about the disaster response and recovery. Finally, we did not identify any clear mechanisms for tracking different rates of recovery within any of the communities that could be related to NPO activities. Consequently, we were unable to assess the evolution and impact of NPO activities over time for different subgroups. Our research was limited by a low response rate from NPOs we contacted about participation in the study. In addition, we were unable to conduct site visits. Thus, future research is needed to provide more insight into recovery tracking. Although this was beyond the scope of our study, NPOs might perform similar functions in the presence of a PDD. However, because PDDs bring more federal resources, it is also likely that victims in those cases are less reliant on NPO aid to help them fully recover from a disaster.

Summary of Key Findings

Qualitative data about recovery processes after nondeclared disaster events reveal several insights about the role of NPOs. First, the number of NPOs reportedly involved in each event varies, although types of activities typically focused on direct humanitarian relief tailored to the sociodemographics of community members' needs. Coordination and communication were enhanced by organizations whose mission focused on these issues (e.g., VOADs, public radio). A lack of federal assistance posed challenges to the efficiency and potential equity of recovery efforts. Limited public information was available about the role of NPOs in longer-term economic recovery.

CHAPTER 4

Roles of Nonprofit Organizations in Disaster Recovery as Discovered Through Grant Data

To complement the qualitative document analyses and interviews described in Chapter 3, we explored a quantitative approach to understand NPOs' role in recovery. Specifically, we sought databases that cataloged nonprofit grant activities, looking for empirical evidence about NPOs' role in long-term recovery following nondeclared disaster events. We anticipated— but did not find—data that could be used to depict the nature of networks among NPOs and between NPOs and community members. Nonetheless, NPO grant data revealed (1) varied profiles of funding across the case studies, (2) a predominance of small grants that occurred largely within the first year following a disaster, and (3) a need for innovative data-collection approaches to provide more-detailed insights about NPO activities during long-term disaster recovery.

Approach to Identifying Grant Data for Nonprofit Organizations

Most databases of nonprofit grants are based on annual Internal Revenue Service (IRS) Form 990 filings, which did not serve our purposes. Candid's Foundation Directory (Candid, undated), however, was designed for general purposes surrounding research and verification of NPOs, finding funding, and exploring research questions about NPO grantmaking and receiving.[1] Candid's data on nonprofits come from IRS information returns, IRS Exempt Organizations Business Master File Extract, individual funders or NPOs, organization web-sites, news, and press releases. Notably, much of the NPO data contained in Candid's data-base come from funders or nonprofits that choose to share information. Hence, Candid does not contain the universe of grantmakers or recipients and provides only a limited view into NPO activity during disaster recovery. This, as well as the general lack of publicly available data on NPO activities during disaster recovery, highlights the need for centralized and com-prehensive data-collection efforts.

---

[1] Candid began in 2019 when the Foundation Center and GuideStar joined forces and can be used for research, including to verify nonprofit information, find funding, and explore questions and issues.

Candid was not specifically designed to locate nonprofit funding tied to specific disasters, but multiple text fields can be used to identify grants on specific topics, including the primary subject of the grant, support strategies employed, and a general description field. These text fields, however, are of inconsistent detail and often missing. Therefore, we used a two-stage approach to identify funding tied to disaster recovery in the five years preceding each disaster:

1. **First, cast a broad net with automated searches.** For each case study, we conducted a broad search that was likely to capture a wide variety of grants, including those that are disaster related.
2. **Second, refine the search results with manual review.** We examined the grant descriptions to select only those that seemed disaster related.

The incompleteness of the text fields means that this is likely an undercount of NPO funding. (See Appendix D for details of the search strategy, including specific search terms.)

Profiles of Nonprofit Grant Funding Vary Across the Case Studies

The grant funding profiles based on Candid data vary across the case studies. Figure 4.1 presents the total amount of funding received by each community from NPOs, as captured in the Candid database. For comparison, we include total funding from non-FEMA federal agencies (e.g., Bureau of Indian Affairs, U.S. Environmental Protection Agency), as well as FEMA, to the extent that these are also captured in the Candid database. One entry, $520,000 in 2016 from the Denali Commission to Newtok, Alaska, was classified as non-FEMA federal because it was directly funded by the federal government.[2] Keeping in mind qualifications about the reliability or validity of Candid data, Figure 4.1 might suggest that an absence of federal assistance does not necessarily lead to more funding from NPOs in the set of case studies selected for this project.

As shown in Figure 4.1, each of the six case studies has a unique funding profile, with four cases (Dumas, Arkansas; Howard County, Indiana; Tampa, Florida; and Dallas, Texas) having exclusively NPO entries. Yarnell has the largest NPO amount, accompanied by substantial FEMA funding. Newtok, Alaska, has no NPO funding, at least as captured in Candid, but a substantial mix of federal funding. And whereas Newtok is the location that filed for a PDD (and hence its selection as a case study), it is difficult to completely isolate Newtok from the surrounding communities that also received money, sometimes in grants that were meant to be split among the villages in the local council. For example, 21 grants are allocated

---

[2] Funded by the Denali Commission Act of 1988 (Pub. L. 105-277, 1998, Title III), the Denali Commission provides grants to rural communities, largely related to infrastructure (Denali Commission, undated).

FIGURE 4.1

Total Grant Funding for Each Case Study, as Captured in the Candid Database

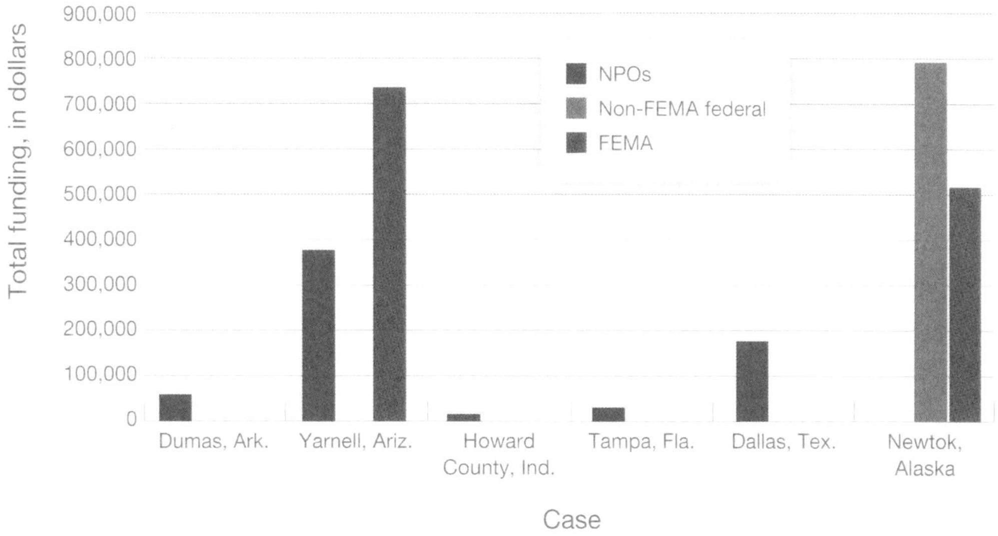

to the village of Tununak, about 30 miles southwest of Newtok, but, in such a sparse area, these funds will inevitably benefit and affect Newtok in some marginal fashion. Other grants that are equally hard to classify are those meant for causes less directly related to disasters (e.g., two grants were sent for the Newtok "storytellers media campaign"). Although such efforts might not directly benefit the transplanting process the village is experiencing, these are necessary efforts to maintain a cultural throughline between Newtok and Mertarvik, the location to which the village will be moving. As detailed in Appendix D, Tampa and Dallas received very large numbers of grants (more than 30,000 each) during the five-year search window, but very few of them mentioned specific disaster-related terms in their descriptions.

Nonprofit Organizations Typically Provide Small Grants Within One Year of a Disaster Event

Many of our case studies received small numbers of grants from NPOs to assist in disaster response and recovery, and these grants were typically awarded in the year a disaster occurred or in the year following the disaster. The fact that the description for each grant in Candid is so limited makes it difficult to determine each grant's specific purpose, but the small amounts and immediate timing suggest that the focus was on humanitarian relief rather than longer-term economic recovery.

Table 4.1 details just the NPO grants for each case study, resulting from the above two-stage approach, including grantmaker name, recipient name, year authorized, grant amount, an overall description of the grant, and the number of years after the disaster in which the

TABLE 4.1
Detailed List of Grants Related to Disaster, from Candid Data

Location	Grantmaker Name	Recipient Name	Year	Grant Amount, in Dollars	Description	Years After Disaster
Dumas, Arkansas	Arkansas Community Foundation, Inc.	Dumas [Public] School District	2008	4,000	To YAC and VICA Med for Disaster Response Team for High School and Jr. High	1
		First United Methodist Church Dumas Arkansas	2008	1,750	For funeral expenses for storm victim who died after rebuilding house	1
		City of Dumas	2007	50,000	To assist city with disaster relief after tornado	0
Yarnell, Arizona	Arizona Community Foundation	Yarnell Community Center	2014	400,100	Yarnell hill recovery group - rebuilding Yarnell	1
Howard County, Indiana	Church Mutual Insurance Company Foundation. Inc.	Kokomo Urban Outreach Inc	2016	10,000	To assist with community recovery after Kokomo EF3 tornado	3
	Old National Bank Foundation Inc	United Way of Howard County	2016	5,000	Disaster Relief for Kokomo Indiana	3
	The Allstate Foundation		2014	1,000	General Operating Support	1
Tampa, Florida	Rooms to Go Children's Fund	American Red Cross	2016	25,000	Disaster relief for Louisiana flood victims	1
			2015	5,000	Disaster relief for South Carolina flood victims	0
Dallas, Texas	Community Foundation of North Louisiana	Retreet	2019	35,000	Replant trees in Ruston following the tornado	0
	The Meadows Foundation	American Red Cross (Dallas)	2019	25,000	Toward providing disaster relief to tornado victims in Ellis, Dallas, Collin and Rockwall Counties	0
		Texas Baptist Men	2019	25,000	Toward disaster relief services to tornado victims in Dallas and surrounding areas	0

Table 4.1—Continued

Location	Grantmaker Name	Recipient Name	Year	Grant Amount, in Dollars	Description	Years After Disaster
Dallas, Texas, continued	The Ginger Murchison Foundation	Preston Hollow Presbyterian Church Foundation	2019	20,000	Tornado Relief; Scholarship Programs	0
	TREC Community Investors	Dallas Independent School District	2019	15,000	DISD Tornado Relief donation	0
	Harvey R., Jr. and Patricia W. Houck Foundation, Inc.	Dallas Foundation	2019	12,500	Tornado disaster relief	0
	Sam Roosth Foundation	Temple Emanu-El	2019	10,000	Tornado relief	0
	Dallas Jewish Community Foundation	American Red Cross (Dallas)	2019	8,750	Disaster relief	0
	Walter & Olivia Kiebach Foundation	Tip of the Spear Foundation	2019	5,000	General program support	0
	The Marilyn Augur Family Foundation	The Cambridge School of Dallas	2019	5,000	For tornado damage	0
			2019	5,000	For tornado damage	0
			2019	5,000	For tornado damage	0
	The Pinkston Foundation	American Red Cross - Dallas Fort Worth Chapter	2019	3,500	Disaster relief donation	0
	Campbell Foundation	American Red Cross (Dallas)	2019	2,500	For disaster relief in the North Texas area (after tornados in Dallas October 2019)	0
	Community Foundation of North Texas	United Way	2019	1,000	For children and families in need effected by tornados in Dallas county	0
	The Jeffrey A. Carter Foundation	American Red Cross - Dallas Fort Worth Chapter	2019	250	Provide funds for emergency disaster relief	0

Table 4.1—Continued

Location	Grantmaker Name	Recipient Name	Year	Grant Amount, in Dollars	Description	Years After Disaster
Dallas, Texas, continued	Aetna Foundation, Inc.	American Heart Association, Inc.	2019	25[a]	Disaster Response	0
Newtok, Alaska	None					

NOTE: Abbreviations are as they appear in the data. We have not provided definitions here because we cannot be certain what was intended.

[a] Although $25 is the entry in the database, we wonder whether perhaps this is a reporting error.

grant was awarded. Typically, grantmakers are community foundations (government grants captured in Candid are detailed in Appendix D). Recipients were generally schools, churches, city governments, fire departments, foundations, and other NPOs. Grant amounts were generally small but ranged from $25 to $400,100.[3] Dumas, Arkansas, received three grants totaling $55,750; Yarnell, Arizona, received one grant from an NPO totaling $400,100; Howard County, Indiana, received three grants totaling $16,000; Newtok, Alaska, received no grants from NPOs but did receive substantial funding from federal grants (see Table D.2 in Appendix D); Tampa, Florida, received two grants totaling $30,000; and Dallas, Texas, received 17 grants totaling $178,525 to assist in disaster recovery. As noted above, we emphasize that these dollar amounts are lower-bound estimates of the amounts received, because of the often-limited and inconsistent information in the text data fields.

The sixth column in Table 4.1 provides a brief description of the who, what, or why the funds were awarded. Examples of why the funds were awarded include to help with disaster relief after a tornado, funeral expenses for a storm victim, assistance to firefighters, general support, predisaster mitigation, children and families in need after storm, and disaster relief.[4] The final column shows that most of our case studies received funds in the year a disaster happened or one year later. In contrast, Howard County, Indiana, received funds for disaster relief three years after the disaster occurred.

It is worth noting that these results were culled from much larger sets of results from the initial searches (stage 1 of our strategy above). Particularly in the cases of the larger cities (Tampa, Florida, and Dallas, Texas), there was extremely high volume of nonprofit grant activity, including for other disasters in the area (e.g., Hurricane Irma for Tampa, Florida, in 2017). It is entirely possible that some of that activity would benefit disaster recovery.[5] Of note is that, whereas there was relatively little grant activity specifically referencing the village of Newtok, Alaska, there was an extremely large amount of activity in the surrounding county, which had received significant press regarding climate change. Presumably, Newtok would benefit from some of these regional resources.

---

[3] We suspect that that the very low value of $25 could be an error in data entry by the original data reporters, which illustrates the challenges resulting from using the self-reported data in Candid.

[4] Because text was brief and often imprecise, we consider grants that might include aspects of both response (e.g., immediately assisting with food, water, or other emergency supplies) and recovery (e.g., providing longer-term resources for rebuilding houses or schools).

[5] As detailed in Appendix D, because of the volume of grants identified for Tampa and Dallas, we used additional search terms to narrow the topic of the grants to be more disaster specific.

Innovative Approaches to Data Collection Are Needed to Provide Insight About Nonprofit Organizations' Activities During Long-Term Disaster Recovery

The limitations of existing data underscore a need for new data-collection methods that provide insight into NPO activities during long-term disaster recovery. For example, FEMA's Individual and Community Preparedness Division (ICPD) runs the Post Disaster Survivor Preparedness Research Survey. This product allows ICPD, after a PDD, to conduct interviews, focus groups, or surveys of communities affected by the disaster. This could, for example, be a set of interviews with NPO representatives regarding their roles in disaster recovery, where their funds come from, and where their resources are expended. Such an endeavor could provide great insight into what an NPO is doing during disaster recovery, as well as the network of NPO actors. One challenge to using this data-collection mechanism, however, is the requirement for a PDD. To better understand cases in which the federal government is not involved, it might also be beneficial to collect these data during nondeclared events.

Another example of an innovative approach might be to explore possible partnerships between federal agencies and NPOs, such as Candid, as one of the few existing data sources. The analysis described in this report was based on publicly available data from the Candid database, with a variety of limitations noted. However, Candid undoubtedly has access to more-extensive data, with greater flexibility for data pulls and analysis. Candid runs a variety of subscription services, which provide varying access to the data. These subscriptions might be beyond the means of many underserved communities. DHS might consider a partnership with Candid, perhaps as a superuser with a license to help underserved communities access and benefit from these existing data. Such a partnership might also help Candid better understand the needs of such communities, especially within the context of long-term disaster recovery.

Summary of Key Findings

Within the Candid data, grants specifically related to our case studies were few and varied widely in their scope and funding levels. Nonprofit grantmaking was most prevalent in the immediate aftermath of the disaster, perhaps indicating that these organizations were filling a role of supplying funding to address short-term needs rather than longer-term recovery. The mix of NPO versus federal (FEMA and non-FEMA) varied widely across cases, with a greater amount of NPO funding within the more-urban settings. Findings related to Newtok, Alaska, suggest a need to capture regional funding, especially when focusing on small communities that might be better characterized (at least in part) as part of a larger region.

The clearest finding from our examination of the Candid database is a general unavailability of centralized data on NPO activities in disaster recovery. IRS filings are the most robust data source, which is summarized in several databases but does not specify activities at the granularity needed for the current analysis. Candid data on NPO grantmaking was the single best source and likely an incomplete one at that. Specifically, it has the following limitations, which might be instructive of general challenges to collecting such data systematically:

- Candid relies on self-reports from funders or nonprofits that choose to share information, which makes it likely that data are missing (if an organization does not self-report) or that data entry was unreliable (if, for example, the $25 Aetna Foundation grant in Dallas was likely missing a few zeros).
- The fact that the description of each grant in Candid is limited makes it difficult to determine a grant's specific purpose. Hence, the data might not capture all aspects of funding, such as richness in location, disaster timing, and activities supported.
- The ability to identify disaster-related grants via manual review varies between urban and rural areas. For example, large urban areas receive a very large volume of grants, which makes detailed manual review less feasible.
- Candid tracks only those NPO efforts for which a grant is issued. Many recovery efforts rely on other funding mechanisms.

Even Candid in ideal form would capture only grants issued by NPOs, which is clearly a small slice of the total role NPOs must play in disaster recovery. Commercial entities (e.g., Home Depot, Walmart) might rely on corporate-affiliated foundations or in-kind contributions that do not use formal grant mechanisms. Similarly, religious organizations and other nonprofits might use their normal operating budgets to pivot to disaster recovery. Tracking these resources is considerably more challenging than tracking grants made.

CHAPTER 5

Metrics for Assessing the Effectiveness of Nonprofit Organizations in Disaster Recovery

Measuring NPO effectiveness in disaster recovery is key to understanding and leveraging NPO capabilities. In this chapter, we describe metrics related to four main categories: (1) community recovery over time, (2) the quality of disaster recovery processes, (3) changes in community-level social capital, and (4) NPO performance capabilities. Measures in each of these categories are needed to ensure a comprehensive assessment of NPO roles and impact within the disaster-affected social system.

Several criteria guided selection of the metrics:[1]

- First, metrics need to be **measurable against a predisaster baseline** so, as conditions change in a community after a disaster, with or without NPO support, metrics used to assess recovery can track these changing conditions.
- Second, metrics need to be **simple**, and the information presented needs to be clear and easily understood by a variety of audiences.
- Third, metrics need to be **accessible** and rely on existing data rather than requiring new data collection to avoid stretching limited resources, especially in underserved communities.
- Fourth, metrics need to **balance a short- and long-term focus** because recovery occurs over many years and the type of information needed will evolve.
- Finally, metrics need to be **reliable over time**—that is, sensitive to long-term changes (e.g., metrics that pertain to ecosystem health should incorporate data on slow-moving, climate-driven hazards).

At least two categories of metrics are of interest: *predisaster* metrics that could be used to anticipate NPO effectiveness and, in turn, successful recovery and *postdisaster* metrics to track NPO-related recovery. In this chapter, we focus on metrics that capture the strengths of NPOs that make them well suited to tackle specific areas of postdisaster needs. For instance,

---

[1] These criteria are adapted from Horney et al. (2017); Horney et al. developed their criteria by drawing on best practices from the fields of sustainable development and health indicators. Given that the focus of this report is on community resilience, we do not emphasize the criterion related to individual experiences.

the social capital brought by NPOs make them particularly well suited to tackle such aspects of disaster recovery as

- Physical & mental health care
- Short & long-term housing
- Employment
- Case management
- Other social services
- Donations & volunteer management
- Animal care
- Spiritual & emotional care. (Acosta and Chandra, 2013, p. 363)

In support of predisaster metrics, the literature review in Chapter 2 identified several roles that NPOs do or could play in disaster recovery, what NPOs need to be successful, and steps that communities and the federal government could take to help bolster NPO effectiveness. A key takeaway from that chapter was that communities and government institutions can bolster or dampen NPO effectiveness based on their own attributes and contributions (see, for instance, Islam and Walkerden, 2015). Consequently, predisaster metrics that anticipate NPO effectiveness need to take a composite or systemic view of NPOs, the communities they are intended to serve, and the regulatory and policy landscape within which NPOs are expected to operate. The contents of Table 2.1 in Chapter 2 can help guide the development of such metrics.[2]

The extant literature on *postdisaster* recovery metrics is vast but falls short in a few ways. First, the literature tends to place primary focus on built infrastructure, sidelining social or economic aspects of recovery, although there is a growing awareness of the need to consider social aspects of recovery (Chandra and Acosta, 2009; Narayanan et al., 2020). Second, not all metrics are easily tied to specific disasters and the recovery that follows because, for instance, of a lack of baseline data. Finally, the existing literature provided little to no guidance on how to track aspects of disaster recovery that might specifically follow from NPO involvement—for instance, some aspects of recovery of a community that received both local or federal assistance and NPO support might relate to government activities, NPO activities, or a combination of both.

Considering these limitations, **the sampling of metrics we present in this chapter is intended to serve as a starting point** for communities and NPOs to develop customized sets of metrics that suit their specific, contextual needs. We strove to identify metrics that embody at least a few of the five criteria listed above. Although we do not explicitly note the equity implications of each, **differences between populations (defined by socioeconomic status or other demographic features) would ideally be captured in the measurement of all metrics presented in this chapter.** For instance, assessments of the extent to which housing condi-

---

[2] RAND's ENGAGED toolkit (Acosta, Chandra, Towe, et al., 2016a) provides a structure framework that NPOs and those aiming to evaluate them can use to assess their anticipated effectiveness.

tions have improved postdisaster should, where possible, consider how different populations (particularly those that have traditionally been marginalized) fare in this regard. Moreover, to the extent that there are significant population shifts in a community, evaluations of recovery should consider progress made both by the new entrants and by the groups they might have displaced.

Metrics for Tracking Nonprofit Organizations' Effectiveness in Disaster Recovery Processes

Anchoring on social and economic domains of recovery and drawing from a broad set of recovery metrics (Abramson, Redlener, et al., 2010; Acosta and Chandra, 2013; Horney et al., 2017), we identified a sample set of metrics that might be particularly relevant for assessing NPO effectiveness in communities where they have been involved. The metrics included in Table 5.1 are related to one of four broad areas of disaster recovery that NPOs are poised to address. The most-relevant metrics for any given context will certainly vary according to characteristics of the NPO, community, and disaster event.

As with any metric that is intended to track change over time (in this case, improvement along some recovery dimension following a disaster), lack of a baseline renders meaningless any snapshot aimed at capturing change in the health of a system (in this case, a community after a disaster). In selecting the example metrics we present in this chapter, we favored those that were likely to rely on data that are refreshed periodically, allowing continued, long-term comparisons of pre- and postdisaster conditions.

Across contexts, metrics that are relevant for tracking a given NPO's effectiveness in a community following a particular disaster ought to be tied to specific activities that the NPO undertook in that community in support of recovery efforts for that disaster. As noted in Chapter 4, such record-keeping is not common practice at present, at least not in a centralized manner. Even the best available source is incomplete, relies on organizations' self-reporting of activities, and is intended to catalog only NPO-issued grants, which likely represent a small portion of NPOs' total contributions to disaster recovery. Similarly, interviews conducted as part of the case studies described in Chapter 3 revealed that NPO-related recovery outcomes are often not tracked systematically.

For stakeholders to examine how well communities are capitalizing on opportunities afforded by NPOs, metrics need to focus on the quality of the disaster recovery process. Example metrics related to the process are described in Table 5.2.

To ensure a comprehensive assessment of NPOs' impact on community-level social capital, stakeholders need metrics that capture involvement in community networks. An example is the use of data from Mediamark Research about social capital at the county level (Hawes and Rocha, 2011). The data come from a stratified sample of more than 20,000 people inter-

TABLE 5.1

Example Metrics for Tracking Nonprofit Organizations' Effectiveness in Disaster Recovery

Domain	Example Metric
Physical and mental health	Percentage of community health care facilities that are operational
	Percentage of NPOs that offer disaster-related medical or mental health support
Housing	Percentage of population who reside in temporary housing units or shared spaces
	Percentage of residential units that are vacant
	Percentage of the population who have been displaced by disaster
	Percentage of housing units that have been abandoned
	Average level of insurance taken out by affected households (total value of homeowner's insurance policies divided by the value of all housing units)
Employment	Unemployment rate
	Percentage of NPOs that offer workforce assistance programs
	Percentage of NPOs that provide vocational education and training
	Annual net income inflow (U.S. dollars per 1,000 residents)
Social services	Percentage of public services (e.g., fire department, law enforcement) that are operational
	Percentage of the eligible population who are registered to vote
	Percentage of protected natural areas that have been restored
	Percentage of damaged cultural arts and religious facilities that have been reconstructed or repaired

SOURCES: Acosta and Chandra, 2013; Horney et al., 2017.

viewed twice a year in the contiguous United States.[3] The index can be used to construct measures of social capital, including community organizational life, engagement in public affairs, and community volunteerism. Example metrics of community social capital are described in Table 5.3.

Finally, a variety of performance metrics is needed to quantify NPOs' capabilities. Performance measurement has gained increased importance in the NPO sector, and numerous frameworks are available (Lee and Nowell, 2015). In Table 5.4, we list example metrics relevant to NPO capabilities during disaster recovery. Metrics address three domains: (1) outcomes, which are the extent to which an NPO satisfies the needs of the population being served; (2) public-value accomplishments, which is the extent to which an NPO creates a valued community; and (3) network or institutional legitimacy, which is the extent to which

---

[3] The contiguous United States excludes Alaska and Hawaii and all other offshore areas, including territories, such as Guam and Puerto Rico.

TABLE 5.2

Example Metrics for Assessing the Quality of the Disaster Recovery Process

Domain	Example Metric
Improved planning	Number and nature of invitees, attendees, and active participants in meetings (conducted by, e.g., NPOs, community members) to enhance community recovery initiatives
	Number and nature of initiatives to improve recovery outcomes specifically for underserved populations
Improved resilience	Number and nature of efforts to institute regulations that enhance disaster resilience (e.g., strengthening building codes)
Expanded outreach methods	Types and instances of media used to engage the public during disaster recovery planning (e.g., brochures; surveys; websites; print, radio, and television advertisements; workshops)
	Type of organizations involved in recovery processes, including charities, community groups, religious entities, and school associations

SOURCE: Horney et al., 2017.

TABLE 5.3

Example Metrics for Assessing Community Social Capital

Domain	Example Metric
Community organizational life	Number of NPOs per 1,000 population
	Number of civic and social organizations per 1,000 population
Engagement in public affairs	Number of people engaged in fund-raising
	Number of people attending public meetings
	Number of people writing or phoning radio or television stations
Community volunteerism	Number and amount of contributions to public radio or television
	Number of people actively volunteering for nonpolitical organizations

SOURCE: Hawes and Rocha, 2011.

the NPO maintains positive relationships with other organizations, reputational legitimacy within the community, compliance with laws, and best practices. Data for these metrics could be obtained through formal, independent evaluation processes; systematic postdisaster assessments; or self-administered surveys.

Summary of Key Findings

Metrics of NPOs' effectiveness in disaster recovery need to capture (1) changes in a community's health, social, and economic functioning; (2) the quality of the disaster recovery process; (3) aspects of the community's social capital assumed to be responsible for NPO

TABLE 5.4

Example Dimensions for Assessing Nonprofit Organizations' Capabilities During Disaster Recovery

Domain	Example Dimension
Outcomes	Level of client or customer satisfaction
	Extent of new clients or customers acquired
Public value accomplishments	Extent of social inclusion
	Extent of individual engagement
Network or institutional legitimacy	Extent of funder relations and diversification
	Level of credibility with other civil society actors
	Extent of compliance with general or particular laws

SOURCE: Lee and Nowell, 2015.

success; and (4) performance along dimensions of NPO capabilities. Notably, NPOs cannot be effective on their own. They need the support of communities and the government (at all levels) to be effective. Although we have identified sample metrics that could shed light on NPO effectiveness across contexts, additional metrics tailored to the unique characteristics of individual communities and disaster events could provide a more comprehensive assessment.

CHAPTER 6

Summary of Findings and Recommendations

In this chapter, we return to the research questions about NPOs' role in long-term recovery in the absence of federal disaster assistance, innovations that enhance NPOs' role in underserved communities, and measures for capturing the effectiveness of NPO support. To address these questions, we began by reviewing existing literature on NPOs' role in disaster recovery. We then examined six case studies of nondeclared natural disasters from diverse communities and geographies. We explored publicly available datasets on NPO grantmaking and identified example metrics for assessing NPOs' unique capabilities that are expected to enhance community capacity for socioeconomic recovery.

Linking Findings to Recommendations

Drawing on the findings reported in each prior chapter, we now explore actions that can be taken to address next steps in DHS's efforts to enhance long-term disaster recovery, including considerations of factors hindering recovery in underserved communities. These actions are grouped as near- and longer-term recommendations. We anticipate that the near-term recommendations could be implemented within the next couple of years, that there would be relatively few barriers to implementation, that change would be primarily within the control of DHS, and that DHS works closely with NPOs. For the longer-term recommendations, we acknowledge that implementation could occur on a longer timeline (more than two years) and require substantial coordination with entities outside DHS.

Table 6.1 summarizes the findings derived from the research described in Chapters 2 through 5 and links them with short- and long-term recommendations. Each recommendation is explained in more detail below, and we close the chapter with some concluding thoughts.

TABLE 6.1

Linking Findings to Recommendations and Potential Primary Actors

			Stakeholder with a Primary Role in Addressing the Recommendation			
			Nongovernmental		Government	
Finding	Term	Recommendation	NPO	Private Sector	State or Local	Federal
NPOs play a key role in disaster recovery, but the benefits within or across communities might not be evenly distributed, and this particularly disadvantages underserved populations.						
	Near	Improve coordination between government agencies and the nonprofit sector to enhance disaster recovery efforts in underserved communities.	x		x	x
	Longer	Develop guidance (including equity considerations) on how to enhance NPO roles across the disaster cycle.	x			x
Qualitative data are lacking about the mechanisms by which NPOs enhance equitable, long-term economic recovery after nondeclared disasters.						
	Near	To more thoroughly assess NPO roles during recovery, follow up with any community that is denied a PDD.	x	x	x	x
	Longer	Identify a set of communities (including underserved populations) and metrics to track NPO roles and impacts in long-term socioeconomic recovery.	x		x	x
Comprehensive, centralized data on NPO activities in disaster recovery are generally unavailable.						
	Near	Develop a conceptual framework for NPOs' roles in disaster recovery, and use this to prioritize data collection.	x		x	x
	Longer	Foster external partnerships with data-gathering organizations to advance data collection (e.g., determining priorities for survey items) and dissemination.	x	x	x	x
Metrics of NPO effectiveness in disaster recovery need to capture multiple dimensions (socioeconomic outcomes, recovery processes, community social capital, NPO capabilities, and equity in processes and outcomes).						
	Near	Explore options internal to DHS for gathering and organizing contextualized information about NPO roles in disaster recovery.	x			x
	Longer	Develop a disaster recovery tracking tool, including metrics related to NPO activities and impacts and community context.	x			x

Near-Term Recommendations

Improve Coordination Between Government Agencies and the Nonprofit Sector to Enhance Disaster Recovery Efforts in Underserved Communities

Best practices for improving coordination between public and nonprofit sectors typically include defining common outcomes, agreeing on roles and responsibilities, and identifying necessary resources (U.S. Government Accountability Office, undated). Such practices, however, might be overlooked in the context of nondeclared disasters because, by definition, involvement by at least one part of the public sector (the federal government) is limited. Soliciting and integrating input about improving coordination from the perspective of NPOs working with underserved communities is particularly important to ensure that appropriate resources (e.g., staff, funding, tools) are identified and context-sensitive nuances (e.g., historical economic or environmental injustices) relevant to implementation are addressed. Additionally, evaluating the success of efforts to improve coordination will be crucial for determining the extent to which lessons learned can be generalized from one community context to another. The public and nonprofit sectors will need to work collaboratively to address this recommendation.

To More Thoroughly Assess Nonprofit Organizations' Roles During Recovery, Follow Up with Any Community That Is Denied a Presidential Disaster Declaration

A systematic and comprehensive qualitative assessment of responses by communities that were denied PDDs is needed to track NPO roles and impacts over time. FEMA could resource such assessments, but data collection and analysis would need to be conducted by a third, impartial party. This includes remaining in contact with local government officials and NPO leaders to discuss their recovery strategies and how they addressed any challenges faced. Establishing mechanisms that support local officials (e.g., emergency managers) or NPOs to document lessons learned in the aftermath of disasters could provide valuable sources of information. Assessing perspectives from diverse stakeholders (e.g., city emergency managers and NPOs) will be important for capturing different experiences and understandings. A sampling strategy could be determined in advance to ensure appropriate coverage of communities representing diverse natural hazard types, geographic contexts, and sociodemographic characteristics. A semistructured interview protocol also could be designed in advance to ensure that data are collected systematically about key topics, such as significant process innovations or examples of when different types of social capital were effective or not and with whom. Conducting interviews with NPO representatives and community members about the intended or expected roles of NPOs, factors that facilitate or impede these roles, and observed or perceived effectiveness will be essential for understanding how communities recover.

Develop a Conceptual Framework for Nonprofit Organizations' Roles in Disaster Recovery, and Use This to Prioritize Data Collection

A detailed conceptual framework and a more robust and reliable data infrastructure are needed for systematic and comprehensive quantitative assessment of NPO roles and impacts following nondeclared disasters. The roles that NPOs play in disaster recovery are manifold, diverse, and largely independent of the federal government. Improved coordination with this important sector has the potential to increase the effectiveness of federal and nonfederal efforts. However, improvement generally requires measurement. Knowing what to measure requires a clear conceptualization of moving parts and their relationships, informed by NPO and community feedback. A conceptual framework could form the basis for defining the domain of measurement, as well as key inputs, processes, outputs, and outcomes. In addition to identifying which NPOs are active in disaster recovery in the long term, data could capture NPO roles beyond grantmaking, how those activities get funded (for example, through in-kind contributions or normal operations), and policies or other factors that create disparities in the uptake of NPO support.

Explore Options Internal to the U.S. Department of Homeland Security for Gathering and Organizing Information About Nonprofit Organizations' Roles in Disaster Recovery

DHS—specifically, FEMA—has a large footprint in disaster recovery, with a vast capacity to collect and organize data. An example provided in Chapter 4 is the Post Disaster Survivor Preparedness Research Survey run by ICPD. This flexible data-collection mechanism could be leveraged to capture NPO activities (grantmaking and otherwise) following disasters. Other such mechanisms and opportunities within DHS might also provide short-term solutions. Casting a wide net during data-collection efforts is important for capturing different perspectives (e.g., across government agencies, private-sector businesses, or NPOs).

Longer-Term Recommendations

Develop Guidance (Including Equity Considerations) on How to Enhance Nonprofit Organizations' Roles Across the Disaster Cycle

In the long term, better guidance is needed to enhance NPO roles across the disaster cycle. NPO and community perspectives need to be central to informing development of this guidance to ensure that NPO missions, community contexts, and real-world equity considerations are addressed. The complex and dynamic disaster environment requires a deep understanding of NPOs' roles in the disaster management system and the fundamental role that networks and different types of social capital play in ensuring community resilience and recovery. A key to NPOs' ability to provide specialized and effective solutions is their knowledge of local communities' needs and dynamics. Developing guidance on how to successfully leverage the important and innovative roles of NPOs will be best done outside disaster situations. Although DHS could lead this effort, significant collaboration with NPOs will be needed.

Identify a Set of Communities (Including Underserved Populations) and Metrics to Track Nonprofit Organizations' Roles and Impacts in Long-Term Socioeconomic Recovery

Commitment to a long-term, mixed-method effort is needed to understand the complex nature of disaster recovery and the mechanisms by which nongovernmental actors, such as NPOs, effectively and equitably tackle the social challenges being faced by disaster-affected communities with limited federal resources. An interdisciplinary, multilevel analysis of the role and impact of social capital in disaster recovery will need diverse types of data and analytic approaches. Building a study design with input from end users of the information, including affected communities, will help to identify priority research questions and knowledge gaps about when and why the benefits of social capital are unevenly spread and how to reduce existing disparities. Collaborative partnerships between DHS, research institutions, and NPOs will be needed to identify a set of communities and relevant metrics for longitudinal research that provides an in-depth understanding of real-world recovery processes, particularly in underserved areas. These partnerships will also be valuable for identifying optimal frequencies and modalities for sharing lessons learned, to enhance long-term recovery with best practices used by NPOs.

Foster External Partnerships with Data-Gathering Organizations to Advance Data Collection (e.g., Determining Priorities for Survey Items) and Dissemination

Partnerships with external organizations (e.g., Candid) will be valuable in the long term to advance data-collection and dissemination efforts. Currently, data collection and cataloging by external organizations are not optimized for understanding NPO effectiveness or equity performance in long-term disaster recovery, but such a purpose could help shape their processes. Potential users of these data include government agencies at all levels, NPO leadership, communities, and disaster researchers and practitioners. Partnerships among these stakeholders could also help to facilitate access to data for underserved communities that might not be able to afford Candid subscriptions yet need to contribute information or access specific outputs to track disparities over time.

Develop a Disaster Recovery Tracking Tool, Including Metrics Related to Nonprofit Organizations' Activities and Impacts and Community Context

A disaster recovery tracking tool would be helpful for a community regardless of whether it receives a PDD. This tool could focus specifically on (1) level of impact, including socioeconomic impacts; (2) local government recovery strategies; (3) NPO recovery strategies; and (4) community context (including dimensions of social capital or census variables). Ideally, communities could use this tracking mechanism to document their recovery at several time intervals in order to fully assess NPO response effectiveness in the long term. Notably, however, local assessments might be subject to reporting bias if they are tied to federal resources, so this potential challenge would need to be addressed. DHS and FEMA could examine the data to identify effective and equitable recovery strategies and distribute this information to local governments and NPOs to assist with disaster preparedness.

Ideally, evaluation metrics are tailored to specific contexts, where context is defined by the community, disaster, and specific activities that the NPO undertook in support of recovery. Using such metrics requires that NPO activities and outcomes be systematically tracked by the NPOs themselves, by the communities they serve, or both. Predisaster baseline measures will be important to collect also to ensure that change over time can be assessed. DHS could help to create mechanisms and provide resources for centrally storing, maintaining, and disseminating such information to address the need for knowledge in varied, evolving, disaster recovery contexts.

Concluding Thoughts

This work provides the DHS Science and Technology Directorate with an assessment of what is known about NPO roles in community recovery following nondeclared disasters based on peer-reviewed literature, case studies, and grantmaking data. In short, knowledge gaps are large, and data are sparse. To assist in future efforts to enhance long-term recovery for communities after nondeclared disasters, we identified various sets of metrics aimed at capturing health, social, and economic functioning; the quality of recovery processes; dimensions of social capital; and NPO capabilities likely associated with better or worse outcomes. Immediate next steps to address knowledge gaps include robustly capturing the perspectives of NPOs and community residents on their experiences of disaster recovery without significant federal support. Codeveloping study designs and interview protocols with NPOs and members of affected communities from a variety of geographic and sociodemographic contexts will help ensure that useful (i.e., credible, relevant, valid) information is generated for rapid uptake by diverse decision makers to support improvements in disaster recovery processes.

The context-sensitive and evolving nature of disaster recovery means that no single study or data source will be able to answer all questions. Rather, an integrated data system that can be accessed by a portfolio of projects with different aims, methods, and time horizons is needed. In the short term, DHS might be able to capitalize on existing resources to identify and analyze the necessary data but will need to develop—in close collaboration with NPOs—a conceptual framework to help prioritize new data collection. In the longer term, multimethod, longitudinal research will be needed, along with new partnerships and tools aimed at understanding and supporting disaster recovery, especially among the most-vulnerable members of disaster-affected communities.

Case-Study Literature Review: Detailed Methodology and Findings

To identify relevant sources for this study, we conducted a systematic literature review. A RAND librarian conducted a keyword search using relevant databases and identified an initial set of sources. Table A.1 provides a list of keywords.

Databases searched included KS Discovery, PubMed, Scopus, U.S. Newsstream, and Web of Science. In addition, we conducted an advanced Google search to identify additional gray literature. Types of sources identified included academic articles and local and national news articles. After the initial literature search was completed, project researchers used DistillerSR literature review software to determine the relevance of each source and explore how sources

TABLE A.1

Literature Review Keyword Search Terms

Subject	Keyword Set
Yarnell, Arizona, wildfire (2013)	("Yarnell" OR "Yarnell Arizona" OR "Yarnell AZ" OR "Yarnell Hill Fire" OR "Yavapai County Arizona" OR "Yavapai County AZ")
Dumas, Arkansas, tornado (2007)	("Dumas" OR "Dumas Arkansas" OR "Dumas AR" OR "Desha County Arkansas" OR "Desha County AR")
Newtok, Alaska, climate change (ongoing)	("Newtok" OR "Newtok Alaska" OR "Newtok AK" OR "Bethel Census Area" OR "Ningliq River" OR "Mertarvik Alaska" OR "Mertarvik AK")
Howard County, Indiana, tornado (2013)	("Howard County" OR "Howard County Indiana" OR "Howard County IN" OR "Kokomo Indiana" OR "Kokomo IN")
Dallas, Texas, tornado (2019)	("City of Dallas" OR "Dallas" OR "Dallas City" OR "Dallas Texas" OR "Dallas TX" OR "Dallas County Texas" OR "Dallas County TX")
Tampa, Florida, flooding (2015)	("City of Tampa" OR "Tampa" OR "Tampa Florida" OR "Tampa FL" OR "Tampa Bay Florida" OR "Tampa Bay FL")
Type of disaster	("wildfire*" OR "wild fire*" OR "tornado*" OR "flood*" OR "climate change" OR "global warming" OR "greenhouse effect")
Disaster relief	("disaster declaration*" OR "disaster relief" OR "disaster recovery" OR "FEMA")

addressed key questions related to the research project.[1] Table A.1 provides an overview of our review of sources. For **level 1** of the review, we examined the title and abstract of each document to determine whether the source pertained to one of the six case studies. After this was completed, 244 sources were included and 16 were excluded. For **level 2** of the review, we categorized each source by type. At the completion of this level, 242 sources were included and two were excluded because they were determined to be duplicates. For **level 3**, we obtained full-text documents for all remaining sources and answered key questions about each source. Questions included whether the source provided a detailed description of the disaster, what types of data (if any) were used in analysis and discussion, and whether the source discussed NGO involvement. At the completion of this level, 238 sources were included and four were excluded because, after reading the full text, we determined that the source was not relevant to our study.

Table A.2 shows the number of sources that were included and excluded during the level 1 review for each case, and Tables A.3 through A.5 provide details about each level of review by case. The data show that the literature search for Yarnell, Arizona, yielded more sources than the searches for other cases. This can be attributed to substantial media coverage and policy analysis following the deaths of the 19 Granite Mountain Hotshot firefighters. The literature search for Howard County, Indiana, yielded the fewest results. Note that the totals between tables differ based on the phase of the review and the issue on which we focused.

Table A.3 gives a detailed look at our level 2 review, which categorized sources by type. Because of the absence of a PDD for each case, there is not a centralized source of data detailing the response and recovery process. As a result, information was drawn from several different types of sources. The results show that local and national news articles and briefs were the most common sources for all the cases, often providing the most details about the response and recovery processes. Yarnell, Arizona; Tampa, Florida; and Newtok, Alaska, yielded the most peer-reviewed articles. In all three cases, we found that this was because the disasters

TABLE A.2

Number of Documents Included and Excluded at Each Level of Review

Review Level	Included	Excluded
1	244	16
2	242	2 (duplicate sources removed)
3	238	4

---

[1] DistillerSR software supports a systematic review of literature by allowing the user to create a database of sources and analyze the sources for key themes. More information about the program can be found at DistillerSR (undated). Literature searches for each case were conducted from the year of the disaster to the present. For example, for Yarnell, Arizona, we searched 2013 to present. It is important to note that the publication of online information has increased significantly over time. This might have created constraints in identifying information for older cases, particularly the 2007 Dumas, Arkansas, tornado.

were framed as part of larger problems. For example, the deaths of the firefighters in Yarnell, Arizona, led to research on best practices for responding to wildfires and preventing future tragedies. The flood in Tampa, Florida, became part of larger discussions about coastal flood mitigation. Newtok, Alaska, has become a case study for discussing the ongoing impact of climate change.

Tables A.5 and A.6 provide more details about the full-text review, including the phase of the disaster on which the source focused and details about the type of response discussed by the source. Table A.5 shows that the majority of sources were descriptive, providing details about what happened when the natural disaster hit the community. A smaller subset of articles focused on the immediate aftermath and short-term recovery, which included the availability of emergency shelters and food assistance. A total of 43 sources focused on the long-term recovery effort, which included rebuilding, and a total of 31 sources focused on the long-term impact, which included discussions of how the disaster affected the community's future disaster planning.

TABLE A.3

Number of Documents in Level 1 Review, by Case

Case	Included	Excluded	Total
Dumas, Arkansas (2007)	34	8	42
Yarnell, Arizona (2013)	118	4	122
Howard County, Indiana (2013)	12	1	13
Tampa, Florida (2015)	32	1	35
Dallas, Texas (2019)	20	2	22
Newtok, Alaska (ongoing)	26	0	26
Total	242	16	258

TABLE A.4
Number of Documents, by Source Type, for Each Case Study

Source Type	Dumas, Ark. (2007)	Yarnell, Ariz. (2013)	Howard County, Ind. (2013)	Tampa, Fla. (2015)	Dallas, Tex. (2019)	Newtok, Alaska (Ongoing)	Total
Local news articles and briefs	8	57	7	9	7	2	90
National news articles and briefs	19	24	2	2	5	5	57
Magazine articles	2	4	2	0	0	4	12
Policy briefs and reports	0	10	0	3	1	2	16
Conference proceedings and papers	0	4	0	0	0	1	5
Peer-reviewed articles	0	10	0	15	1	8	34
Press releases	3	3	0	1	2	0	9
Legislative documents	2	0	0	0	0	0	2
Editorials	0	3	0	2	1	1	7
Blog posts	0	2	0	0	0	0	2
Websites	0	0	1	0	1	0	2
Other	0	1	0	0	0	3	4
Total	34	118	12	32	18	26	240

TABLE A.5

Number of Documents Identified for Each Phase of Disaster, by Case

Case	Description of Event	Immediate Aftermath	Short-Term Recovery	Long-Term Recovery	Preparedness	Long-Term Impact
Dumas, Arkansas (2007)	17	5	12	4	3	0
Yarnell, Arizona (2013)	37	2	2	12	11	6
Howard County, Indiana (2013)	3	2	3	4	0	0
Tampa, Florida (2015)	5	3	3	5	14	13
Dallas, Texas (2019)	14	12	2	3	2	1
Newtok, Alaska (ongoing)	18	0	0	13	4	12
Total	94	24	22	41	34	32

TABLE A.6

Number of Documents Identified for Disaster Response Types, by Case

Case	Federal Response	State or Local Response	NPO Response	Economic Recovery	Financial Resources for Victims	Material Resources for Victims
Dumas, Arkansas (2007)	12	0	3	0	1	1
Yarnell, Arizona (2013)	13	10	6	1	2	0
Howard County, Indiana (2013)	7	0	2	1	0	0
Tampa, Florida (2015)	12	0	2	0	0	0
Dallas, Texas (2019)	4	11	1	3	1	2
Newtok, Alaska (ongoing)	16	2	4	0	0	0
Total	64	23	17	5	4	3

Key-Informant Interviews: Detailed Methodology and Findings

We interviewed six key informants from our case-study sites to gain more insight into response and recovery processes. To identify key informants, we contacted local and state emergency management officials and leaders of NPOs in each site's jurisdiction. We emailed letters explaining the purpose of our study and inviting them to participate. In total, we contacted 37 potential respondents and six agreed to participate. We faced two primary challenges in securing interviews. First, our ability to identify potential participants varied depending on the transparency of the organization structure of local and state government offices and NPOs. Second, because of the time elapsed between the disasters and our interviews, there was considerable turnover in government and NPO offices. The Dallas city official and non-profit coordinator were the only informants who had been in their positions when the disaster occurred. This made it more difficult to identify key informants and to gather detailed information about the disaster from those who agreed to participate. Table B.1 provides more detail about sampling and key informants.

Each interview lasted between 45 minutes and one hour. We used separate questionnaires for government officials and for NPO coordinators. Questions focused on the impact of the disaster, response strategies, and recovery (see the boxes after Table B.1 for the full lists of questions). When relevant, we also probed about the disaster's impact on underserved communities. Interviews were recorded, and interview transcripts were coded according to the most-relevant themes for our study.

TABLE B.1

Key-Informant Sample

Case	Number of Potential Respondents Contacted	Number of Respondents Interviewed	Key-Informant Description
Dumas, Arkansas (2007)	11	1	State emergency management official
Yarnell, Arizona (2013)	3	1	County emergency management official
Howard County, Indiana (2013)	10	0	Not applicable
Tampa, Florida (2015)	7	2	County emergency management official
			NPO disaster response coordinator
Dallas, Texas (2019)	2	2	City emergency management official
			NPO disaster response coordinator
Newtok, Alaska (ongoing)	4	0	Not applicable
Total	37	6	

Interview Questions for Government Officials

1. Tell me a little bit about what you do for a living.
2. How long have you lived in [community]?
3. Tell me about the [name of disaster].
   a. Where were you living in the community when it happened?
   b. What impact did it have on the community?
   c. How did it compare to other disasters the community has experienced?
4. How prepared was [community] for [name of disaster]?
   a. Were there any specific things in place to deal with [disaster]?
5. How did [community] deal with the immediate aftermath of [disaster]?
6. Were any government, community or volunteer organizations involved in helping with the immediate aftermath?
7. What did they do?
8. What role did you play in dealing with the immediate aftermath [name of disaster]?
9. How has [community] recovered from [disaster]?
   a. How does your community track the recovery? (Probe: Do you have specific metrics? Do you look at any specific issues, such as business closures, state of infrastructure, poverty rate, or provision of social services?)
   b. What are some things that have gone particularly well in the recovery?
   c. What are the most-significant challenges [community] has faced?
   d. What role have you played in helping with the recovery?
10. Which nonprofit organizations helped with disaster recovery?
    a. What role did the organizations play in the recovery effort?
    b. How well did the nonprofit organizations communicate with the community?
    c. What aspects of your community's recovery were overlooked by the nonprofits that were involved? (Probe: What could the organizations have done better?)
11. What resources or innovations could have helped the nonprofit organizations do a better job?
12. What role have nonprofits played in helping your community prepare for future disasters?
13. Is there any assistance that community needed that you did not receive? (Probe: What type?)
14. How could the community's recovery efforts be improved in the future?

Interview Questions for Nonprofit Organizations' Coordinators

1. Tell me a little bit about [name of organization].
   a. What is your current position in the organization?
      i. How long have you worked in this position?
      ii. What are your job responsibilities?
2. Were you working in your current position when [disaster] happened? If no:
   a. What position were you working in?
   b. Were you living in the area?
3. What impact did [disaster] have on [name of community]?
4. How did it compare to other disasters the community has experienced?
5. How prepared was [community] for [name of disaster]?
   a. Were there any specific things in place to deal with [disaster]?
6. How did [community] deal with the immediate aftermath of [disaster]?
7. What role did your organization play in dealing with the immediate aftermath of [name of disaster]?
8. What role did your organization play in the recovery effort?
9. How did you inform community members about your organization's efforts?
   a. Did your organization face in obstacles in informing the community about your efforts?
10. What tools did you use to distribute aid to community residents?
    a. Were there obstacles in distributing aid?
11. Did your organization partner with any other nonprofits to deliver aid?
    a. If yes: How did you coordinate your efforts?
12. What role did insurance companies play in the recovery?
    a. Did your organization work with any insurance companies?
13. Were there parts of the recovery with which your organization was unable to assist?
14. How has [community] recovered from [disaster]?
    a. What are some things that have gone particularly well in the recovery?
    b. What are the most-significant challenges [community] has faced?
15. What impact did your organization's work have on the recovery?
    a. Does your organization track the impact of your efforts on the recovery? (Probe: Do you have specific metrics? Do you look at any specific issues, such as business closures, state of infrastructure, poverty rate, or provision of social services?)
16. How does your organization prepare for dealing with future disasters like [name of disaster]?
    a. Have you made any changes since [name of disaster]?
       i. Has [name of disaster] affected how you prepare for and respond to future disasters?
17. How could the community's recovery efforts be improved in the future?

Descriptive Summary of Case Studies

Case Study 1: Tornadoes in Dumas, Arkansas, 2007

Overview

On February 2007, an F3 tornado hit Dumas, Arkansas (Committee on Homeland Security, 2007). An estimated 40 residents were injured (Squires, 2007). The tornado destroyed the homes of 37 families. In addition, one of the town's major employers, the Arkat Nutrition animal feed plant, was completely destroyed, and several other businesses were damaged significantly. As a result, 800 residents were left unemployed (Brodsky, 2007). Our literature review yielded 34 references about the tornadoes. In addition, we interviewed one key informant, a state emergency management official.

Response and Recovery

Governor Mike Beebe requested an emergency declaration on February 27, 2007. The request was denied on March 8, 2007, because FEMA determined that the state had sufficient resources to manage the response and recovery. However, FEMA did agree to provide 23 mobile homes and seven travel trailers (Committee on Homeland Security, 2007).

News reports indicate that the largest NPO response came from Catholic Charities USA. The organization set up a relief program at a local health center to provide assistance to residents, which included training 50 caseworkers to work with victims. Available assistance included meals, household items, and school supplies (Hargett, 2007).

Challenges

In our literature review, we found that, according to local officials, the lack of federal assistance was the most significant challenge Dumas faced in the response to the tornadoes. On March 15, 2007, Beebe testified before the U.S. House of Representatives Committee on Homeland Security in a hearing on FEMA and PDDs. Beebe said he received off-the-record information from a FEMA official that the state's budget surplus had affected the decision to deny the state's request for a PDD. He told the committee,

> . . . I think consistent with the whole FEMA philosophy and the whole FEMA espoused
> and stated policy, and that is don't punish a state, don't punish a community for helping

themselves. Don't punish people who have a good plan in place to take care of themselves. We don't expect FEMA to solve all of our problems. We don't expect the expect the federal government to solve our problems. We will take care of Arkansans one way or the other, whether we get any federal help or whether we don't get any federal help. (Committee on Homeland Security, 2007, p. 3)

However, he went on to say, "But, it is so much easier and quicker and better and more thorough if we can be a partner with the federal government and obtain that assistance" (Committee on Homeland Security, 2007, p. 3).

Case Study 2: Wildfire in Yarnell, Arizona, 2013

Overview

On June 28, 2013, Yarnell, Arizona, experienced a significant wildfire. The wildfire burned 8,300 acres and destroyed 93 homes. Most significantly, 19 firefighters, known as the Granite Mountain Hotshots, were killed when they became trapped while fighting the fire (FEMA, 2022b). Our literature review yielded 118 references about the fire. In addition, we interviewed a Yavapai County emergency management official.[1]

Response and Recovery

Governor Jan Brewer requested a major disaster declaration on July 9, 2013. The request was denied on August 9, 2013, because FEMA determined that the state had sufficient resources to manage the response and recovery. Brewer appealed the denial, and the appeal was denied on September 13, 2013 (FEMA, 2013).

Although Yarnell was not granted a PDD, there was a significant NPO response. This was in large part due to the deaths of the Granite Mountain Hotshots, which brought national attention to the fire and its impact on the community. Reports of NPO funding included the following:

- An estimated $1,592,903 was raised in charitable donations to help Yarnell residents with the recovery (DeFilippis, 2014). About $636,067 was spent specifically on rebuilding uninsured homes.
- As of June 2014, the 100 Club of Arizona had raised $244.5 million for families of the Granite Mountain Hotshots, and $243.5 million had been distributed. The Phoenix and Prescott firefighter associations raised $248.5 million for families of the Granite Mountain Hotshots, and all of their funds had been distributed (Anglen and Wiles, 2014).
- The Arizona Community Foundation and Yavapai County Community Foundation raised $400,000 from businesses and individuals to assist victims with rebuild-

---

[1] Yarnell is in Yavapai County.

ing homes and other needed resources. In addition, they provided aid to the Yarnell Water Improvement Association and the Yarnell and Peeples Valley fire departments. They also established the Yarnell Memorial Scholarship Endowment, which had raised $460,000 in 2013 (Arizona Community Foundation, 2013).

In addition, Yarnell received federal funding that was not contingent upon a disaster declaration. For example, the U.S. Small Business Administration offered small-business loans of up to $2 million, home loans of up to $200,000, and personal property loans up to $40,000 (FEMA, 2022b). The U.S. Department of Agriculture Rural Development program provided a $40,000 grant to help the Yarnell Fire District rebuild a helipad destroyed in the fire (Rural Development, 2017).

Challenges

Literature on the impact of the fire focuses primarily on challenges that contributed to the deaths of the Granite Mountain Hotshots. Yavapai County utilized the Arizona Disaster Recovery Framework (AZDRF) in its response, which had been developed in 2012.[2] However, analyses of the fire have concluded that the community had not been adequately prepared to manage the wildfire. In addition, emergency response officials did not fully consider weather conditions in their response, which contributed to the deaths of the Granite Mountain Hotshot firefighters. Findings from research on the problems with the response have been used to revise the AZDRF (Santos, 2016; Wilson, 2013).

We interviewed an emergency management official in Yavapai County. Although they were not in the emergency management office at the time of the fire, they had been working in a lower-level emergency response position at the time of the fire and had significant insight into the impact that it had on the community. They told us that response coordination was one of the most-significant challenges. The official explained that, despite the existence of the AZDRF, coordination between the emergency management office and the local police department was challenging. In addition, residents were slow to evacuate, making the rescue and recovery efforts more challenging. The official noted that the AZDRF did not have a mechanism for tracking long-term disaster recovery. As a result, there is not a clear way to measure the effectiveness of the response.

---

[2] The AZDRF is part of the *Arizona State Emergency Recovery and Response Plan* and outlines the roles of state emergency officials to facilitate an organized response in the immediate aftermath of a disaster. The full plan can be accessed at Arizona Department of Emergency and Military Affairs, 2017.

Case Study 3: Tornado in Howard County, Indiana, 2013

Overview

On November 17, 2013, an EF 3 tornado hit Howard County, Indiana. An estimated 32 residents were injured by the tornado and its effects, and thousands were without power for up to a week.[3] Damage to local homes totaled an estimated $2.8 million, and damage to local infrastructure totaled an estimated $309,840. In addition, an estimated 60 local homes and 30 local businesses were completely destroyed (Gerber, 2014). We reviewed 12 references and were unable to secure an interview with a key informant.

Response and Recovery

Governor Mike Pence requested a PDD on December 3, 2013. The request was denied on December 2013 because FEMA determined that the state had sufficient resources to manage the response and recovery. Pence appealed the denial, and the appeal was rejected on January 7, 2014 (FEMA, 2014).

Several NPOs provided relief funds to storm victims, including the following:

- WTHR, Howard County's local public radio station, published a list of resources on November 19, 2013. It directed victims to the Red Cross and the United Way of Howard County. They also directed victims to a temporary resource distribution center where victims could access basic needs, such as bottled water and baby formula.
- A year after the storm, the *Kokomo Tribune* reported that 1,275 people volunteered with the United Way of Howard County's long-term recovery team to help clean up debris. In addition, the team reported collecting $119,688 in donations. Donations were used to assist 109 local families. Victim payments of less than $1,000 did not have restrictions on use, while payments of more than $1,000 were overseen by the recovery team's case managers (Gerber, 2014).

Challenges

Our literature review suggests that, according to local officials and residents, the most significant challenge that Howard County faced following the tornado was the lack of aid from FEMA. The November 2013 tornado also followed significant flooding in Howard County in April 2013 that had caused an estimated $7 million in damage. FEMA had also denied the county's request for a PDD in that case, straining county resources. Thus, when the November tornado hit, the county was already struggling with costs. In addition, although the state

---

[3] The National Weather Service uses the EF scale to measure the strength of a tornado. The scale ranges from EF0 to EF5. An EF3 tornado is described as severe with wind speeds between 136 and 165 miles per hour. More information about the scale can be found at National Weather Service, undated.

had a disaster relief fund offering grants of up to $5,000 to disaster victims, funds had been exhausted by the 2013 flood (Hayden, 2014; Myers, 2017).

Howard County commissioner Paul Wyman expressed concern about the disaster declaration denials in a 2017 *Kokomo Tribune* article:

> "When you're standing amongst the damage, and you're looking at it, it's really hard to believe you don't meet a threshold like that," said Howard County Commissioner Paul Wyman. "When people are totally losing their homes and you're looking around at a neighborhood that has a tremendous amount of devastation, it's very disheartening." (Myers, 2017)

An April 2014 report in the *Kokomo Tribune* profiles a November 2013 tornado victim who struggled to recover from the storm's impact. The exterior and interior of her home were both damaged significantly by a combination of wind and water. As a result, her family had been forced to move into temporary housing. She experienced trouble negotiating with her insurance company and was forced to use personal savings for unmet needs. According to county officials, this victim was one of many who had been negatively affected by the lack of assistance from FEMA (Hayden, 2014).

Case Study 4: Flood in the Tampa Bay Region, Florida, 2015

Overview

Between July 25 and August 3, 2015, Pasco, Hillsborough, and Pinellas counties received between 15 and 20 inches of rain.[4] This was the worst flooding in the area in 65 years and caused significant property damage. An estimated 1,000 homes were damaged, and the cost of damage was estimated to be $2.2 million (Wallace, 2015). Our literature review of the disaster included 34 references. In addition, we interviewed an emergency management official for Pinellas County and an emergency response coordinator for the Red Cross in central Florida. Although neither had been in their position at the time of the 2015 flood, they were able to provide significant insight into disaster response in the area. We reviewed 34 references and interviewed two key informants—a county emergency management official and a Red Cross disaster response coordinator.

Response and Recovery

Governor Rick Scott requested a PDD on August 25, 2015. The request was denied because FEMA determined that the state had sufficient resources to deal with the disaster. Scott

---

[4]   The Tampa Bay region is spread across Pinellas, Hillsborough, Pasco, and Manatee counties (Tampa Bay Regional Planning Council, undated).

appealed the denial on September 11, 2015, and the appeal was denied on September 23, 2015 (FEMA, 2015).

News reports document specific local and NPO response and recovery efforts:

- The Red Cross provided emergency shelters for storm victims (Lush, 2015).
- The Salvation Army provided food assistance for storm victims (Owiye, 2015).
- Volunteers with Samaritan's Purse, a disaster relief organization based in North Carolina, helped with debris removal. This included providing assistance to homeowners whose homes experienced significant damage (Schmidt, 2015).

The flooding also had an impact on the area's long-term storm preparation. For example, the city of Tampa is engaged in ongoing flood mitigation planning because of frequent experiences with flooding and evidence that climate change will increase the severity of flooding (College, undated; Locke, 2021).

Challenges

The Pinellas County emergency management official we interviewed emphasized that they had not been in their current emergency management position in 2015 but knew that the area faced significant obstacles when it was denied a PDD. They explained that it can be difficult to secure adequate financial resources to meet the needs of disaster victims:

> . . . the challenge is, the money is not the same when it's a smaller, more localized event. You know, you don't see the fundraising—and I mentioned that before—you don't have the same federal resources that come in. Sometimes you have to fight to get a declaration, and then a lot of times, which we've had quite often here in Pinellas, you don't get that declaration.

They went on to explain that this creates a problem because "there is no money to offer survivors for rebuilds or their needs because, unless the local nonprofits are raising it, it's just not there."

The emergency management official recalled that the area had not received a disaster declaration after the 2015 flood. When we asked how the county navigates situations like this, the emergency management official candidly explained that the fact that resources are limited means that not all victims will receive adequate assistance:

> . . . it starts simple with things that you just can't do, like case management—how do you uncover unmet needs if you don't have the money to do case management? And if you don't know what the unmet needs are, how can you even seek out additional funding to help meet those needs, right? So it's like a cycle; there's no winning it, right? Like what few little resources you're again local; local folks can pour out, and you try to stretch those. But the reality is, some people just won't recover from something like that.

However, they also told us that Florida might be better prepared to manage natural disasters than many states are, even when limited federal resources are available. They told us that this is because of a high level of coordination between state officials, local governments, and NPOs. This is in part because of a strong state VOAD that provides coordination for nonprofits. They explained that the county emergency management office works closely with the VOAD organization any time there is a disaster. In addition, the VOAD organization has a representative who works directly with the governor's office, which creates a strong relationship between nonprofits and the state government.

The disaster response coordinator for the Red Cross also emphasized the strong disaster response coordination efforts in central Florida. They explained that much of the Red Cross's focus is providing short-term aid to victims in the immediate aftermath of a disaster. This includes providing food and shelter to those whose homes have been damaged or destroyed. However, they also work to aid victims with limited resources until those victims become stable. This could include providing financial assistance for housing, repairs, and replacement of other resources that victims might have lost in the disaster. They also coordinate with other nonprofits in the area to provide aid. However, the response coordinator was not aware of a tracking mechanism that would offer clear insight into exactly how long the Red Cross provides aid to those with limited resources and how much aid it provides.

Case Study 5: Tornado in Dallas, Texas, 2019

Overview

Dallas, Texas was hit by an EF3 tornado on October 20, 2019. The tornado was part of a storm system that also brought a total of ten tornadoes to surrounding communities (with the EF3 in Dallas being the largest). The damage was primarily structural, causing an estimated $2 billion in damage, which was the most expensive disaster in Dallas history. More than one dozen Dallas ISD buildings were damaged, and three had to be demolished (Richter, 2019). We reviewed 26 references on the disaster. In addition, we interviewed an emergency management official for the city of Dallas and a disaster response coordinator for a local nonprofit.

Response and Recovery

Texas Governor Greg Abbott requested a PDD on January 13, 2020. FEMA denied the request on March 31, 2020, because it was determined that Texas had sufficient resources to deal with the disaster (Manuel, 2020). Governor Abbott appealed the denial on April 27, 2020, and FEMA denied the request for appeal on June 10, 2020.

The city emergency management official noted that the immediate response had involved significant coordination by city emergency management to respond to challenges faced by residents. They explained,

> We really needed to have a multiagency, multidisciplinary team. Go out there, and actually work is like a convoy. So we had some mutual-aid groups as well come in from our surrounding jurisdictions that just helped us . . . . But what we ended up doing is pairing, you know, a police officer, you know all of the debris removal equipment with, you know, supervisors, a logistics person. . . .

The official went on to explain,

> We were able to do every single thing that needed to be done to open that street right then, so that, as soon as it was done, there was no one else that we had to call—it was all done in one fell swoop.

They told us that they felt that their response strategy worked well and would be used in the future. The official also told us that the local VOAD organization also enabled coordination with local NPOs.

News reports (Richter, 2019) identified several sources of aid, including the following:

- The Church of Jesus Christ of Latter-Day Saints opened a resource center immediately following the storm to help victims.
- On the Road lending provided financial help to people who lost automobiles, which included down payments and loans for new vehicles.
- U-Haul provided 30 days of free storage to storm victims.
- Mark Cuban donated $1 million to Dallas ISD.

Challenges

The Dallas emergency management official we interviewed told us that the lack of a PDD created some obstacles for the city primarily because of costs associated with repairing local infrastructure. The official told us that the most significant challenge for the city was funding to address damage to public buildings. Specifically, there was significant damage to several older school buildings. Although the buildings were insured, their estimated values were low because of their age. As a result, to fully rebuild, they city had to meet a more modern set of standards, which created a significant expense.

The official told us that the city was able to receive smaller amounts of aid from the U.S. Department of Transportation, which helped with infrastructure repair. In addition, they told us that the most heavily damaged area was a predominantly middle-class neighborhood where most homeowners were insured, which made the recovery more manageable. However, they said that they believed that, if the tornado had hit another area of the city with higher levels of economic disadvantage, it would have been more challenging to assist affected residents. They explained that, because the area that was impacted was of the demographic it was, "we didn't have a lot of people that were needing shelter, things like that. Most folks like use their credit card to go get a hotel room, basically." They went on to tell us,

> You know from it—from an individual homeowner standpoint . . . I mean I think we realized very quickly that virtually every structure impacted was insured. And that, I think, was the really critical difference between whether this happened in Preston Hollow or happened in South Dallas—it would have been a different thing.

The NPO disaster response coordinator we interviewed had a different perspective from the city emergency official's on the impact of the tornado. Their nonprofit was unique because it provided immediate and long-term assistance to disaster victims. They assisted victims until it was determined that the victim was in a stable position, which could take several weeks or months.

Although the Dallas emergency management official emphasized that the recovery was less challenging because the tornado hit a predominantly middle-class neighborhood, the NPO disaster response coordinator said that the impact was much more widespread. Their agency was responsible for helping the entire Dallas metropolitan area, and they emphasized that the totality of the storm system, which included the tornado in Dallas and two in small surrounding communities, was significant. The smaller tornadoes affected more-vulnerable populations, including a trailer park community and a multipurpose space that served as a studio and home to several local artists. The disaster response coordinator told us that the lack of a PDD from FEMA was challenging and created a significant cost burden for their organization:

> You know, they were facing huge repair bills; a lot of them are under insured, so really just having the resources available to everybody who either can't leave because of income won't leave or just really need to repair their property [is what mattered]. And, you know, sometimes you're making really difficult decisions with them of, you know what, "What can [our NPO] can help fix right now that will kind of get you through until either more funding comes available or until you save up some money?" And so those kinds of conversations are really hard, so just more funding to assist especially when it's not, you know, a declared disaster [is what is needed].

Their agency provided long-term assistance to members of these groups. The response coordinator also noted that their agency often had contact with people who were less likely than most to seek aid from local government officials, such as undocumented immigrants. Thus, they often find that even smaller disasters, such as single-building fires, can have a significant impact on vulnerable populations in ways that local government might not fully understand.

Case Study 6: Climate Change Impacts in Newtok, Alaska, Ongoing

Overview

Newtok, Alaska, is one of several Alaska Native villages that has been heavily affected by climate change in the past several decades (Alaska Department of Commerce, Community, and Economic Development, 2011). The small village is struggling to deal with thawing permafrost, flooding, and erosion—issues that have been exacerbated by climate change (Ristroph, 2021; Welch, 2019). Changing conditions have compromised much of the village's critical infrastructure and jeopardized the safety of residents' homes. Research has suggested that much of the town could erode and fall into the local Ninglick River by 2027. As a result, residents are being relocated to Mertarvik, which is approximately 10 miles from Newtok (Welch, 2019). Newtok continues to work toward mitigating the impact of climate change, primarily by relocating residents and securing resources that minimize the impact that the significant disruption has on their well-being. We reviewed 26 references and were unable to secure an interview with a key informant.

Response and Recovery

Newtok Village president Paul Charles requested a major disaster declaration on December 24, 2016, to mitigate climate change's effect on the community. The request focused specifically on the impact of erosion, permafrost degradation, and flooding dating back to January 1, 2006. The request was denied on January 18, 2017, because FEMA determined that the request fell outside of the Stafford Act's definition of *natural disaster*.

Newtok received support from several federal sources before and after the disaster declaration denial:

- In 2006, the Bureau of Indian Affairs Housing Improvement Program funded the construction of three homes in Mertarvik for relocating Newtok residents. These were the first homes built at the Mertarvik site.
- In 2016, Newtok received a $900,000 imminent-threat grant from the U.S. Department of Housing and Urban Development's Indian Community Development Block Grant Program for the relocation of 12 families. Funding was for the construction of new homes and infrastructure in Mertarvik (MacArthur, 2016; U.S. Department of Housing and Urban Development, 2018).
- Newtok has received significant pass-through funding from the Denali Commission, a federal entity established in 1998 to support Indigenous communities in Alaska with economic and infrastructure assistance. In 2018, Congress appropriated $15 million to the commission to assist with Newtok's relocation to Mertarvik. The commission granted much of the funding to the Alaska Native Tribal Health Consortium, which

then distributed the funding to residents to assist with their relocation to Mertarvik (Ristroph, 2021).

- In 2018, Newtok received a $1.7 million grant from FEMA's Hazard Mitigation Grant Program to purchase homes in Newtok that were no longer inhabitable and assist residents with their relocation efforts (Ristroph, 2021).
- The Bureau of Indian Affairs has provided funding to rebuild homes in Mertarvik for Newtok residents who have relocated.
- The U.S. Department of Defense's Innovative Readiness Training program has provided labor support for the construction of roads, homes, and other critical infrastructure (Ristroph, 2021).

In addition, the Alaska State Legislature provided $4 million in 2010 and $2.5 million in 2011 for the construction of the Mertarvik Evacuation Center (Demer, 2016).

Challenges

Newtok is one of the first communities in the United States that might be completely eliminated by 21st-century climate change. The situation is unprecedented, and, as a result, securing adequate resources to navigate the changes and provide relocation options for residents presents ongoing challenges. Although FEMA recognizes the impact of climate change, the effects have not been fully integrated into how major disasters are defined. Newtok's experience reveals the challenges that communities might face when responding to climate change and highlights the need for a more streamlined response in the future.

Nonprofit Organizations' Grant Data: Detailed Methodology and Findings

Method

NPO grant data were accessed from Candid Foundation Center and GuideStar (Candid, undated). The Candid database was designed for general purposes surrounding research and verification of nonprofits, finding funding, and exploring research questions. It was not specifically designed to locate NPO funding tied to specific disasters. That said, multiple text fields can be used to identify grants on specific topics. These text fields, however, are of inconsistent detail and often missing.

Therefore, we used a two-stage approach to identify funding tied to disaster recovery:

1. **First, cast a broad net with automated searches.** For each case study, conduct a broad search that is likely to capture a wide variety of grants, including those that are disaster related.
2. **Second, refine the search results with manual review.** Examine the grant descriptions to select only those that look like they are disaster related.

Because of the incompleteness of the text fields, we emphasize that, for each search, this is likely an undercount of NPO funding.

For the first step, we used Candid's Advanced Search and Filters to filter results related to each case study. Figure D.1 shows an image of Candid's Advanced Search and Filters window. Each search includes two main filters: one specifying location (using the **Geographic Focus** field) and one specifying a five-year search window (using the **Year(s)** slider), dating forward from the year of each focal disaster.[1]

Table D.1 shows our search criteria, by case study. For Dumas, Arkansas; Yarnell, Arizona; and Howard County, Indiana, we searched for all recipients of NPO grants that occurred within their respective five-year search windows. Our fourth case study, Newtok, Alaska, is a

---

[1] Dallas, Texas, is the only location with a three-year search window because its natural disaster occurred in 2019. For Tampa, Florida, and Dallas, Texas, we included additional search criteria in the **Subject Area** search bar. See below for more details.

FIGURE D.1
Candid's Advanced Search and Filters

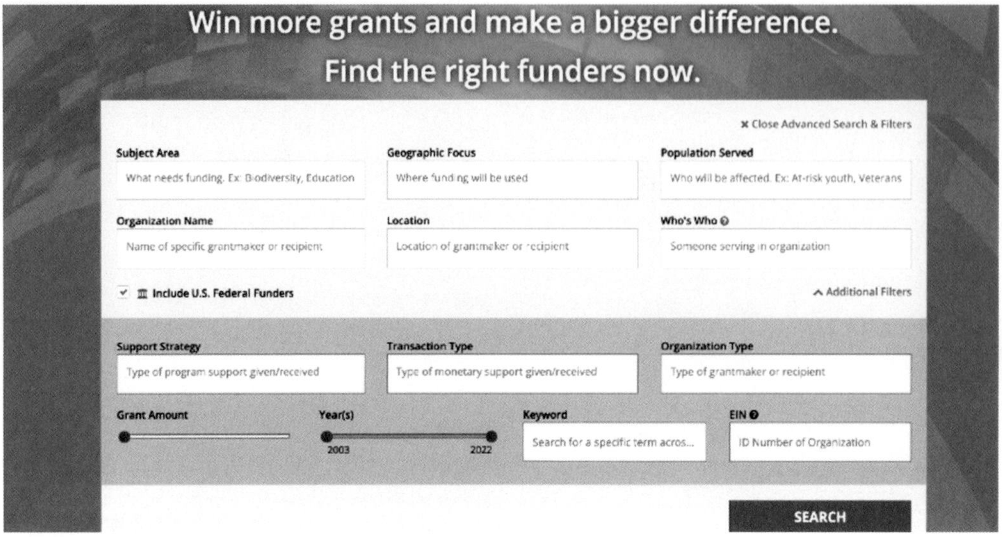

SOURCE: Candid, undated.

small village so is not listed as a searchable location in Candid's advanced search. As a result, we used its corresponding county, Bethel County, as our geographic location criterion. We identified NPO grants designated for Newtok rather than other areas in Bethel County by searching for "Newtok" in the recipient list. For our last two case studies, Tampa, Florida, and Dallas, Texas, we included additional search refinements that were not included in the other case studies for two reasons: (1) Tampa and Dallas are large cities, and (2) each had multiple disasters in its respective time window. Therefore, they had received many NPO grants (i.e., Tampa had received 44,847 grants, and Dallas had received 32,830 grants). Because this first search stage was designed to create a larger set for manual review and to scope down the magnitude of this effort, we included additional search criteria for these two case studies. For each, our search refinements included grants that were related to disaster preparedness, disaster reconstruction, disaster relief, and search and rescue, as well as a term specific to the type of disaster in our case studies (floods for Tampa, Florida, and storm, hurricanes, and tornadoes in Dallas, Texas).[2]

For the second step, we manually identified grant recipients related to our case studies by performing a line-by-line assessment of our yielded search results. The focus here was on specific recipients within a specific location (e.g., Newtok) or likely to be involved in disaster relief (e.g., the American Red Cross), as well as the three text fields that specified the primary

---

[2] Although our focus was on tornadoes in this case study, this search category cannot be separated.

TABLE D.1
Summary of Candid Search Bar Criteria

Geographic Focus	Refinement
Dumas, Arkansas	Recipients located in Dumas, Arkansas (United States)
	Years 2007 to 2012
Yarnell, Arizona	Recipients located in Yarnell, Arizona (United States)
	Years 2013 to 2018
Howard County, Indiana	Recipients located in Howard County (Indiana, United States)
	Years 2013 to 2018
Newtok, Alaska	Recipients located in Bethel Census Area (Alaska, United States)
	Years 2015 to 2020
Tampa, Florida	Recipients located in Tampa (Florida, United States)
	Years 2015 to 2020
	Disaster preparedness
	Disaster reconstruction
	Disaster relief
	Floods
	Search and rescue
Dallas, Texas	Recipients located in Dallas (Texas, United States)
	Years 2019 to 2022
	Disaster preparedness
	Disaster reconstruction
	Disaster relief
	Storms, hurricanes, and tornadoes
	Search and rescue

subject, support strategies, and general description of the funding that were disaster related (e.g., disasters and emergency management).[3]

---

[3] As noted above, in some instances, the description box or subject area was either blank or not informative. In these cases, we could not identify whether the grant was related to disaster recovery.

Supplementary Results

Table D.2 provides detail on the governmental entries in Candid corresponding to each of our case studies.

TABLE D.2

Detail on Governmental Entries in the Candid Database

Location	Grantmaker Name	Recipient Name	Year	Grant Amount, in Dollars	Description	Years After Disaster
Dumas, Arkansas	None					
Yarnell, Arizona	United States Federal Emergency Management Agency	Yarnell Fire Department	2018	316,806	STAFFING FOR ADEQUATE FIRE AND EMERGENCY RESPONSE (SAFER)	5
				58,477	ASSISTANCE TO FIREFIGHTERS GRANT	5
		Yarnell Fire District	2014	368,000	Staffing for adequate fire and emergency response (safer)	1
Howard County, Indiana	None					
Newtok, Alaska	United States Federal Emergency Management Agency	Newtok Village	2018	517,500	PRE-DISASTER MITIGATION	3
	United States Bureau Of Indian Affairs			149,736	ADAPTATION PLANNING (INC DATA DEVELOPM)	3
	Denali Commission		2016	520,000	Environmentally Threatened Communities FY2016	1
	United States Environmental Protection Agency			125,000	Newtok Village will continue to build capacity within their IGAP program, through administrative and environmental processes. Newtok will work on re [entry truncated]	1
			2015	124,175	Tribe will build capacity to develop and maintain environmental programs, by chairing planning committee meetings on the environmental impacts and ne [entry truncated]	0

Table D.2—Continued

Location	Grantmaker Name	Recipient Name	Year	Grant Amount, in Dollars	Description	Years After Disaster
Tampa, Florida	None					
Dallas, Texas	Community Foundation of North Louisiana	Retreet	2019	35,000	Replant trees in Ruston following the tornado	0
	The Meadows Foundation	American Red Cross (Dallas)		25,000	Toward providing disaster relief to tornado victims in Ellis, Dallas, Collin, and Rockwall Counties	0
		Texas Baptist Men		25,000	Toward disaster relief services to tornado victims in Dallas and surrounding areas	0
	The Ginger Murchison Foundation	Preston Hollow Presbyterian Church Foundation		20,000	Tornado Relief [and] Scholarship Programs	0
	TREC Community Investors	Dallas Independent School District		15,000	DISD Tornado Relief donation	0
	Harvey R., Jr. and Patricia W. Houck Foundation, Inc.	Dallas Foundation		12,500	Tornado disaster relief	0
	Sam Roosth Foundation	Temple Emanu-El		10,000	Tornado relief	0
	Dallas Jewish Community Foundation	American Red Cross (Dallas)		8,750	Disaster relief	0
	Walter & Olivia Kiebach Foundation	Tip of the Spear Foundation		5,000	General program support	0
	The Marilyn Augur Family Foundation	The Cambridge School of Dallas		5,000	For Tornado Damage	0
				5,000		0
				5,000		0

70

Table D.2—Continued

Location	Grantmaker Name	Recipient Name	Year	Grant Amount, in Dollars	Description	Years After Disaster
Dallas, Texas, continued	The Pinkston Foundation	American Red Cross—Dallas Fort Worth Chapter	2019, continued	3,500	Disaster Relief Donation	0
	Campbell Foundation	American Red Cross (Dallas)		2,500	For disaster relief in the north Texas area (after tornados in Dallas [in] October 2019)	0
	Community Foundation of North Texas	United Way		1,000	For children and families in need effected by tornados in Dallas county	0
	The Jeffrey A. Carter Foundation	American Red Cross - Dallas Fort Worth Chapter		250	Provide funds for emergency disaster	0
	Aetna Foundation, Inc.	American Heart Association, Inc.		25[a]	Disaster Response	0

NOTE: Abbreviations are as they appear in the data. We have not provided definitions here because we cannot be certain what was intended.

[a] Although $25 is the entry in the database, we wonder whether perhaps this is a reporting error.

Definitions of Key Terms

equity. Fairness related to how benefits and costs are distributed, who is recognized and included in decision processes, and existing conditions that influence access to opportunities and resources (McDermott, Mahanty, and Schreckenberg, 2013).

freedmen's town. A settlement founded between 1865 and 1920 that provided superior economic, political, educational, cultural, or other opportunities and agency for Black Americans (Hunter and Robinson, 2018; Roberts and Matos, 2022).

underserved community. A population systematically denied full access to economic, social, and civic life; includes populations underserved because of geographic location, religion, sexual orientation, gender identity, minority racial and ethnic populations, and special needs (e.g., language barriers, disabilities, alienage status, age) (Biden, 2021; U.S. Code, Title 34, Section 12291).

Abbreviations

AZDRF	Arizona Disaster Recovery Framework
DHS	U.S. Department of Homeland Security
Dallas ISD	Dallas Independent School District
DRRA	Disaster Recovery Reform Act
EF	Enhanced Fujita
FEMA	Federal Emergency Management Agency
ICPD	Individual and Community Preparedness Division
IRS	Internal Revenue Service
NGO	nongovernmental organization
NPO	nonprofit organization
PDD	presidential disaster declaration
VOAD	voluntary organization active in disaster

References

Abramson, David M., Lynn M. Grattan, Brian Mayer, Craig E. Colten, Farah A. Arosemena, Ariane Rung, and Maureen Lichtveld, "The Resilience Activation Framework: A Conceptual Model of How Access to Social Resources Promotes Adaptation and Rapid Recovery in Post-Disaster Settings," *Journal of Behavioral Health Services and Research*, Vol. 42, No. 1, January 2015.

Abramson, David M., Irwin E. Redlener, Tasha Stehling-Ariza, Jonathan Sury, Akilah N. Banister, and Yoon Soo Park, *Impact on Children and Families of the Deepwater Horizon Oil Spill: Preliminary Findings of the Coastal Population Impact Study*, Columbia University, Earth Institute, National Center for Disaster Preparedness, research brief, August 10, 2010.

Acosta, Joie, and Anita Chandra, "Harnessing a Community for Sustainable Disaster Response and Recovery: An Operational Model for Integrating Nongovernmental Organizations," *Disaster Medicine and Public Health Preparedness*, Vol. 7, No. 4, August 2013.

Acosta, Joie D., Anita Chandra, Vivian L. Towe, Yandong Zhao, and Yangxu Lu, *ENGAGED Toolkit: Improving the Role of Nongovernmental Organizations in Disaster Response and Recovery*, RAND Corporation, TL-202-FF, 2016a. As of February 5, 2023:
https://www.rand.org/pubs/tools/TL202.html

Acosta, Joie D., Anita Chandra, Vivian L. Towe, Yandong Zhao, and Yangxu Lu, *The Development and Pilot Testing of the ENGAGED Toolkit*, RAND Corporation, TL-202/1, 2016b. As of January 4, 2023:
https://www.rand.org/pubs/tools/TL202.html

Adie, Charles E., *Holistic Disaster Recovery: Ideas for Building Local Sustainability After a Natural Disaster*, Diane Publishing, 2001.

Alaska Department of Commerce, Community, and Economic Development; Division of Community and Regional Affairs, *Relocation Report: Newtok to Mertarvik*, 2011.

Aldrich, Daniel P., "The Power of People: Social Capital's Role in Recovery from the 1995 Kobe Earthquake," *Natural Hazards*, Vol. 56, No. 3, March 2011.

Aldrich, Daniel P., *Building Resilience: Social Capital in Post-Disaster Recovery*, University of Chicago Press, 2012.

Aldrich, Daniel P., and Kevin Crook, "Strong Civil Society as a Double-Edged Sword: Siting Trailers in Post-Katrina New Orleans," *Political Research Quarterly*, Vol. 61, No. 3, 2008.

Aldrich, Daniel P., Michelle A. Meyer, and Courtney M. Page-Tan, "Social Capital and Natural Hazards Governance," in Susan L. Cutter, ed., *Oxford Research Encyclopedia: Natural Hazard Science*, Oxford University Press, 2018.

Andrews, Rhys, "Social Capital and Public Service Performance: A Review of the Evidence," *Public Policy and Administration*, Vol. 27, No. 1, January 2012.

Anglen, Robert, and Russ Wiles, "Yarnell Charitable Donations Mostly Distributed," *The Republic*, June 28, 2014.

Arizona Community Foundation, "$400,000 Awarded to Help Rebuild Yarnell After Summer Fire," press release, November 13, 2013.

Arizona Department of Emergency and Military Affairs, *Arizona State Emergency Response and Recovery Plan*, February 8, 2017.

Austin, James E., *The Collaboration Challenge: How Nonprofits and Businesses Succeed Through Strategic Alliances*, John Wiley and Sons, 2010.

Ballesteros, Luis, and Aline Gatignon, "The Relative Value of Firm and Nonprofit Experience: Tackling Large-Scale Social Issues Across Institutional Contexts," *Strategic Management Journal*, Vol. 40, No. 4, April 2019.

Bankoff, Greg, and Dorothea Hilhorst, "The Politics of Risk in the Philippines: Comparing State and NGO Perceptions of Disaster Management," *Disasters*, Vol. 33, No. 4, October 2009.

Baregheh, Anahita, Jennifer Rowley, and Sally Sambrook, "Towards a Multidisciplinary Definition of Innovation," *Management Decision*, Vol. 47, No. 8, 2009.

Behera, J. K., "Role of Social Capital in Disaster Risk Management: A Theoretical Perspective in Special Reference to Odisha, India," *International Journal of Environmental Science and Technology*, 2021.

Bhandari, Humnath, and Kumi Yasunobu, "What Is Social Capital? A Comprehensive Review of the Concept," *Asian Journal of Social Science*, Vol. 37, No. 3, January 2009.

Biden, Joseph R., Jr., "Executive Order *13985 of January 20, 2021:* Advancing Racial Equity and Support for Underserved Communities Through the Federal Government," *Federal Register*, Vol. 86, No. 14, January 25, 2021.

Brodsky, Robert, "High and Dry," *Government Executive*, Vol. 39, No. 8, May 15, 2007.

Butler, David, "Focusing Events in the Early Twentieth Century: A Hurricane, Two Earthquakes, and a Pandemic," in Claire B. Rubin, ed., *Emergency Management: The American Experience*, Routledge, 2019.

Candid, Foundation Directory, database, undated.

Chamlee-Wright, Emily, and Virgil Henry Storr, "Social Capital as Collective Narratives and Post-Disaster Community Recovery," *Sociological Review*, Vol. 59, No. 2, May 2011.

Chandra, Anita, and Joie D. Acosta, *The Role of Nongovernmental Organizations in Long-Term Human Recovery After Disaster: Reflections from Louisiana Four Years After Hurricane Katrina*, RAND Corporation, OP-277-RC, 2009. As of January 4, 2023:
https://www.rand.org/pubs/occasional_papers/OP277.html

Chandra, Anita, and Joie D. Acosta, "Disaster Recovery Also Involves Human Recovery," *JAMA*, Vol. 304, No. 14, 2010.

Chandra, Anita, Joie D. Acosta, Stefanie Howard, Lori Uscher-Pines, Malcolm V. Williams, Douglas Yeung, Jeffrey Garnett, and Lisa S. Meredith, "Building Community Resilience to Disasters: A Way Forward to Enhance National Health Security," *RAND Health Quarterly*, Vol. 1, No. 1, 2011. As of January 4, 2023:
https://www.rand.org/pubs/periodicals/health-quarterly/issues/v1/n1/06.html

College, Shawn, team leader, environmental planning, infrastructure and research, Hillsborough County City–County Planning Commission, "TBRPC Peril of Flood Workshop: City of Tampa 2015 Peril of Flood Act Vulnerability Assessment," briefing slides, undated.

Committee on Homeland Security, U.S. House of Representatives, *Disaster Declarations: Where Is FEMA in a Time of Need?* 110th Congress, 1st Session, March 15, 2007.

Curnin, Steven, and Danielle O'Hara, "Nonprofit and Public Sector Interorganizational Collaboration in Disaster Recovery: Lessons from the Field," *Nonprofit Management and Leadership*, Vol. 30, No. 2, Winter 2019.

Cutter, Susan L., "The Vulnerability of Science and the Science of Vulnerability," *Annals of the Association of American Geographers*, Vol. 93, No. 1, March 2003.

DeFilippis, Julianne, "A Year Later, Yarnell Continues Putting a Community Back Together," *Cronkite News*, June 27, 2014.

Demer, Lisa, "The Creep of Climate Change," *Anchorage Daily News*, August 29, 2015, updated November 29, 2016.

Denali Commission, webpage, undated. As of January 5, 2023:
https://www.denali.gov/

DHS—*See* U.S. Department of Homeland Security.

DistillerSR, "DistillerSR: Literature Review Software," webpage, undated. As of January 5, 2023:
https://www.distillersr.com/products/distillersr-systematic-review-software

Dover, Graham, and Thomas B. Lawrence, "The Role of Power in Nonprofit Innovation," *Nonprofit and Voluntary Sector Quarterly*, Vol. 41, No. 6, December 2012.

DRRA—*See* Public Law 115-254, 2018.

Eide, Arne H., "Community-Based Rehabilitation in Post-Conflict and Emergency Situations," in Erin Martz, ed., *Trauma Rehabilitation After War and Conflict*, Springer, 2010.

Eller, Warren, Brian J. Gerber, and Lauren E. Branch, "Voluntary Nonprofit Organizations and Disaster Management: Identifying the Nature of Inter-Sector Coordination and Collaboration in Disaster Service Assistance Provision," *Risk, Hazards and Crisis in Public Policy*, Vol. 6, No. 2, June 2015.

Eller, Warren S., Brian J. Gerber, and Scott E. Robinson, "Nonprofit Organizations and Community Disaster Recovery: Assessing the Value and Impact of Intersector Collaboration," *Natural Hazards Review*, Vol. 19, No. 1, February 2018.

Executive Office of the President, *The Federal Response to Hurricane Katrina: Lessons Learned*, Government Printing Office, 2006.

Fafchamps, Marcel, "Development and Social Capital," *Journal of Development Studies*, Vol. 42, No. 7, 2006.

Federal Emergency Management Agency, U.S. Department of Homeland Security, *National Disaster Recovery Framework: Strengthening Disaster Recovery for the Nation*, September 2011a.

Federal Emergency Management Agency, U.S. Department of Homeland Security, *A Whole Community Approach to Emergency Management: Principles, Themes, and Pathways for Action*, FDOC 104-008-1, December 2011b.

Federal Emergency Management Agency, U.S. Department of Homeland Security, "Arizona—Yarnell Hill Fire: Denial of Appeal," September 13, 2013.

Federal Emergency Management Agency, U.S. Department of Homeland Security, "Indiana—Severe Storms, Straight-Line Winds, and Tornadoes: Denial of Appeal," January 7, 2014.

Federal Emergency Management Agency, U.S. Department of Homeland Security, "Florida—Severe Storms and Flooding: Denial of Appeal," September 23, 2015.

Federal Emergency Management Agency, U.S. Department of Homeland Security, "How a Disaster Gets Declared," webpage, last updated January 4, 2022a. As of January 10, 2023:
https://www.fema.gov/disaster/how-declared

Federal Emergency Management Agency, U.S. Department of Homeland Security, "Yarnell, Arizona Wildfire Recovery: A Coordinated Network of Recovery Support," tool for practitioners, last updated February 16, 2022b.

FEMA—*See* Federal Emergency Management Agency.

Finucane, Melissa L., Joie Acosta, Amanda Wicker, and Katie Whipkey, "Short-Term Solutions to a Long-Term Challenge: Rethinking Disaster Recovery Planning to Reduce Vulnerabilities and Inequities," *International Journal of Environmental Research and Public Health*, Vol. 17, No. 2, January 2020.

Gerber, Kelly Lafferty, "Tornadoes Damage 300 Homes and Commercial Properties," *Kokomo Tribune*, November 17, 2014.

Hargett, Malea, "Catholic Charities Reacts to Dumas Tornadoes: Diocese Training 50 Caseworkers This Week," *Arkansas Catholic*, March 10, 2007.

Hawes, Daniel P., and Rene R. Rocha, "Social Capital, Racial Diversity, and Equity: Evaluating the Determinants of Equity in the United States," *Political Research Quarterly*, Vol. 64, No. 4, December 2011.

Hayden, Maureen, "FEMA's Disaster Decisions Frustrate State, Local Leaders," *Kokomo Tribune*, April 23, 2014.

Hero, Rodney E., "Social Capital and Racial Inequality in America," *Perspectives on Politics*, Vol. 1, No. 1, March 2003a.

Hero, Rodney, "Multiple Theoretical Traditions in American Politics and Racial Policy Inequality," *Political Research Quarterly*, Vol. 56, No. 4, December 2003b.

Hero, Rodney E., *Racial Diversity and Social Capital: Equality and Community in America*, Cambridge University Press, 2007.

Horney, Jennifer, Caroline Dwyer, Meghan Aminto, Philip Berke, and Gavin Smith, "Developing Indicators to Measure Post-Disaster Community Recovery in the United States," *Disasters*, Vol. 41, No. 1, January 2017.

Howell, Junia, and James R. Elliott, "As Disaster Costs Rise, So Does Inequality," *Socius*, Vol. 4, 2018.

Howell, Junia, and James R. Elliott, "Damages Done: The Longitudinal Impacts of Natural Hazards on Wealth Inequality in the United States," *Social Problems*, Vol. 66, No. 3, August 2019.

Hunter, Marcus Anthony, and Zandria F. Robinson, *Chocolate Cities: The Black Map of American Life*, University of California Press, 2018.

Islam, Rabiul, and Greg Walkerden, "How Do Links Between Households and NGOs Promote Disaster Resilience and Recovery? A Case Study of Linking Social Networks on the Bangladeshi Coast," *Natural Hazards*, Vol. 78, September 2015.

Jaskyte, Kristina, "Board Effectiveness and Innovation in Nonprofit Organizations," *Human Service Organizations: Management, Leadership and Governance*, Vol. 41, No. 5, 2017.

Joshi, Abhay, and Misa Aoki, "The Role of Social Capital and Public Policy in Disaster Recovery: A Case Study of Tamil Nadu State, India," *International Journal of Disaster Risk Reduction*, Vol. 7, March 2014.

Joshi, Pamela, *Faith-Based and Community Organizations' Participation in Emergency Preparedness and Response Activities*, Institute for Homeland Security Solutions, February 2010.

Kilby, Patrick, "The Strength of Networks: The Local NGO Response to the Tsunami in India," *Disasters*, Vol. 32, No. 1, March 2008.

Lee, Chongmyoung, and Branda Nowell, "A Framework for Assessing the Performance of Nonprofit Organizations," *American Journal of Evaluation*, Vol. 36, No. 3, September 2015.

Locke, Hannah, "Will Tampa Be the Next Underwater City? Stormwater Management in Tampa, FL," Urban Water Atlas, December 12, 2021.

Lush, Tamara, "Heavy Rain, Flooding Snarl Life in the Tampa Bay Area," *Orlando Sentinel*, August 3, 2015.

MacArthur, Anna Rose, "Federal Grant Helps Newtok Village Relocate Due to Erosion of Ningliq River," Alaska Public Media, May 19, 2016.

Macinko, J., and B. Starfield, "The Utility of Social Capital in Research on Health Determinants," *Milbank Quarterly*, Vol. 79, No. 3, 2001.

Manuel, Obed, "FEMA Denies Texas Appeal for Disaster Relief for October Tornadoes That Caused Millions in Damage in Dallas County," *Dallas Morning News*, June 11, 2020.

Markhvida, Maryia, Brian Walsh, Stephane Hallegatte, and Jack Baker, "Quantification of Disaster Impacts Through Household Well-Being Losses," *Nature Sustainability*, Vol. 3, July 2020.

McDermott, Melanie, Sango Mahanty, and Kate Schreckenberg, "Examining Equity: A Multidimensional Framework for Assessing Equity in Payments for Ecosystem Services," *Environmental Science and Policy*, Vol. 33, November 2013.

McDonald, Robert E., "An Investigation of Innovation in Nonprofit Organizations: The Role of Organizational Mission," *Nonprofit and Voluntary Sector Quarterly*, Vol. 36, No. 2, June 2007.

Moore, Michele-Lee, Frances R. Westley, and Tim Brodhead, "Social Finance Intermediaries and Social Innovation," *Journal of Social Entrepreneurship*, Vol. 3, No. 2, October 2012.

Morris, Susannah, "Defining the Nonprofit Sector: Some Lessons from History," *Voluntas*, Vol. 11, March 2000.

Muller, Alan, and Gail Whiteman, "Exploring the Geography of Corporate Philanthropic Disaster Response: A Study of Fortune Global 500 Firms," *Journal of Business Ethics*, Vol. 84, No. 4, February 2009.

Myers, George, "Kokomo, Howard County Officials Look Back, Express Frustration About FEMA Denials," *Kokomo Tribune*, October 15, 2017.

Narayanan, Anu, Melissa Finucane, Joie Acosta, and Amanda Wicker, "From Awareness to Action: Accounting for Infrastructure Interdependencies in Disaster Response and Recovery Planning," *GeoHealth*, Vol. 4, No. 8, August 2020.

National Centers for Environmental Information, National Oceanic and Atmospheric Administration, "Billion-Dollar Weather and Climate Disasters," webpage, c. 2022. As of September 12, 2022:
https://www.ncei.noaa.gov/access/billions/

National Weather Service, National Oceanic and Atmospheric Administration, "Enhanced Fujita Scale," webpage, undated. As of January 5, 2023:
https://www.weather.gov/tae/ef_scale

Nelson, K. S., and M. Molloy, "Differential Disadvantages in the Distribution of Federal Aid Across Three Decades of Voluntary Buyouts in the United States," *Global Environmental Change*, Vol. 68, May 2021.

Norris, Fran H., Matthew J. Friedman, Patricia J. Watson, Christopher M. Byrne, Eolia Diaz, and Krzysztof Kaniasty, "60,000 Disaster Victims Speak, Part I: An Empirical Review of the Empirical Literature, 1981–2001," *Psychiatry*, Vol. 65, No. 3, Fall 2002.

O'Brien, Karen L., and Robin M. Leichenko, "Double Exposure: Assessing the Impacts of Climate Change Within the Context of Economic Globalization," *Global Environmental Change*, Vol. 10, No. 3, October 2000.

Owiye, Christi, "Salvation Army on Standby for Flood Victims," WGCU, August 4, 2015.

Paldam, Martin, "Social Capital: One or Many? Definition and Measurement," *Journal of Economic Surveys*, Vol. 14, No. 5, December 2000.

Parker, Andrew M., Amanda F. Edelman, Katherine G. Carman, and Melissa L. Finucane, "On the Need for Prospective Disaster Survey Panels," *Disaster Medicine and Public Health Preparedness*, Vol. 14, No. 3, June 2020.

Parks, Vanessa, Lynsay Ayer, Rajeev Ramchand, and Melissa L. Finucane, "Disaster Experience, Social Capitals, and Behavioral Health," *Natural Hazards*, Vol. 104, October 2020.

Post-Katrina Emergency Management Reform Act—*See* Public Law 109-295, 2006.

Public Law 93-288, Disaster Relief Act of 1974, May 22, 1974.

Public Law 100-707, Disaster Relief and Emergency Assistance Amendments of 1988, November 23, 1988.

Public Law 105-277, an act making omnibus consolidated and emergency appropriations for the fiscal year ending September 30, 1999, and for other purposes, October 21, 1998.

Public Law 107-296, Homeland Security Act of 2002, November 25, 2002.

Public Law 109-295, Department of Homeland Security Appropriations Act, October 4, 2006.

Public Law 113-2, an act making supplemental appropriations for the fiscal year ending September 30, 2013, to improve and streamline disaster assistance for Hurricane Sandy, and for other purposes, January 29, 2013.

Public Law 115-254, FAA Reauthorization Act of 2018, October 5, 2018.

Putland, Christine, Fran Baum, Anna Ziersch, Kathy Arthurson, and Dorota Pomagalska, "Enabling Pathways to Health Equity: Developing a Framework for Implementing Social Capital in Practice," *BMC Public Health*, Vol. 13, 2013.

Putnam, Robert D., *Bowling Alone: The Collapse and Revival of American Community*, Simon and Schuster, 2000.

Richter, Marice, "Dallas Tornadoes Caused $2 Billion in Damages," Reform Austin, October 30, 2019.

Ristroph, Elizaveta Barrett, "Navigating Climate Change Adaptation Assistance for Communities: A Case Study of Newtok Village, Alaska," *Journal of Environmental Studies and Sciences*, Vol. 11, September 2021.

Rivera, Danielle Zoe, Bradleigh Jenkins, and Rebecca Randolph, "Procedural Vulnerability and Its Effects on Equitable Post-Disaster Recovery in Low-Income Communities," *Journal of the American Planning Association*, Vol. 88, No. 2, 2022.

Roberts, Andrea, and Melina Matos, "Adaptive Liminality: Bridging and Bonding Social Capital Between Urban and Rural Black Meccas," *Journal of Urban Affairs*, Vol. 44, No. 6, 2022.

Roque, Anaís Delilah, David Pijawka, and Amber Wutich, "The Role of Social Capital in Resiliency: Disaster Recovery in Puerto Rico," *Risk, Hazards and Crisis in Public Policy*, Vol. 11, No. 2, June 2020.

Rothstein, Bo, and Dietlind Stolle, "The State and Social Capital: An Institutional Theory of Generalized Trust," *Comparative Politics*, Vol. 40, No. 4, July 2008.

Runyan, Rodney C., "Small Business in the Face of Crisis: Identifying Barriers to Recovery from a Natural Disaster," *Journal of Contingencies and Crisis Management*, Vol. 14, No. 1, March 2006.

Rural Development, U.S. Department of Agriculture, "USDA Rural Development Helps Yarnell in Its Recovery," press release, December 4, 2017.

Sandy Recovery Improvement Act—*See* Public Law 113-2, 2013.

Santos, Fernanda, "Wildfire Awakens the Sorrow of an Arizona Town Still Scarred by Loss," *New York Times*, June 10, 2016.

Schmidt, S., "Kindness Crests: Relief Groups Do What They Can to Help Pasco County Flood Victims Clean Up the Mess," *Tampa Bay Times*, August 18, 2015.

Simo, Gloria, and Angela L. Bies, "The Role of Nonprofits in Disaster Response: An Expanded Model of Cross-Sector Collaboration," *Public Administration Review*, Vol. 67, December 2007.

Sokolowski, S. Wojciech, "Innovation, Professional Interests, and Nonprofit Organizations: The Case of Health Care in Poland," *Nonprofit Management and Leadership*, Vol. 8, No. 4, Summer 1998.

Squires, Chase, "7 Deaths Are Blamed on Plains Storms," *Daily Breeze*, February 25, 2007.

Szreter, Simon, and Michael Woolcock, "Health by Association? Social Capital, Social Theory, and the Political Economy of Public Health," *International Journal of Epidemiology*, Vol. 33, No. 4, August 2004.

Tampa Bay Regional Planning Council, homepage, undated. As of January 20, 2023:
https://www.tbrpc.org/

Telford, John, Margaret Arnold, and Alberto Harth, *Learning Lessons from Disaster Recovery: The Case of Honduras*, World Bank, Working Paper 35567, June 1, 2004.

Tierney, Kathleen, "Social Inequality, Hazards, and Disasters," in Ronald J. Daniels, Donald F. Kettl, and Howard Kunreuther, eds., *On Risk and Disaster: Lessons from Hurricane Katrina*, University of Pennsylvania Press, 2006.

U.S. Census Bureau, "QuickFacts: Dallas City, Texas," webpage, undated-a. As of January 20, 2023:
https://www.census.gov/quickfacts/fact/table/dallascitytexas/PST045222

U.S. Census Bureau, "QuickFacts: Howard County, Indiana," webpage, undated-b. As of January 20, 2023:
https://www.census.gov/quickfacts/fact/table/howardcountyindiana/PST045222

U.S. Census Bureau, "QuickFacts: Tampa City, Florida," webpage, undated-c. As of January 20, 2023:
https://www.census.gov/quickfacts/fact/table/tampacityflorida/PST045222

U.S. Code, Title 6, Domestic Security; Chapter 1, Homeland Security Organization; Subchapter III, Science and Technology in Support of Homeland Security; Section 185, Federally Funded Research and Development Centers.

U.S. Code, Title 34, Crime Control and Law Enforcement; Subtitle I, Comprehensive Acts; Chapter 121, Violent Crime Control and Law Enforcement; Subchapter III, Violence Against Women; Section *12291, Definitions and Grant Provisions.*

U.S. Department of Homeland Security, "Natural Disasters," webpage, last updated October 5, 2022. As of January 5, 2023:
https://www.dhs.gov/natural-disasters

U.S. Department of Housing and Urban Development, "Safer Ground," webpage, c. May 20, 2016, archived January 8, 2018. As of January 4, 2023:
https://archives.hud.gov/local/ak/goodstories/2016-05-20.cfm

U.S. Government Accountability Office, "Leading Practices in Collaboration Across Governments, Nonprofits, and the Private Sector," webpage, undated. As of October 30, 2022:
https://www.gao.gov/
leading-practices-collaboration-across-governments%2C-nonprofits%2C-and-private-sector

Villalonga-Olives, Ester, and Ichiro Kawachi, "The Measurement of Social Capital," *Gaceta Sanitaria*, Vol. 29, No. 1, January–February 2015.

Wallace, Jeremy, "Obama Rejects Scott's Plea for Disaster Aid," *Tampa Bay Times*, September 4, 2015.

Welch, Craig, "Climate Change Has Finally Caught Up to This Alaska Village," *National Geographic*, October 22, 2019.

Wilson, Reid, "Report Details How Wildfire Overran Arizona 'Hotshots,' Killing 19; Communications Cited," *Washington Post*, September 28, 2013.

Wisner, Ben, Piers Blaikie, Terry Cannon, and Ian Davis, *At Risk: Natural Hazards, People's Vulnerability and Disasters*, Routledge, 2004.

World Population Review, "Dumas, Arkansas Population 2023," webpage, undated-a. As of January 20, 2023:
https://worldpopulationreview.com/us-cities/dumas-ar-population

World Population Review, "Newtok, Alaska Population 2023," webpage, undated-b. As of January 20, 2023:
https://worldpopulationreview.com/us-cities/newtok-ak-population

World Population Review, "Yarnell, Arizona Population 2023," webpage, undated-c. As of January 20, 2023:
https://worldpopulationreview.com/us-cities/yarnell-az-population

Zhang, Qiaoyun, "Disaster Response and Recovery: Aid and Social Change," *Annals of Anthropological Practice*, Vol. 40, No. 1, May 2016.